The diversity of life

FOUNDATIONS OF BIOLOGY

General editor L. M. J. Kramer

A major advanced biology course for schools and colleges

Books in the series are:

The diversity of life

The cell concept

Heredity, development and evolution

Metabolism, movement and control

Man and the ecosystem

The diversity of life

C. Mary Jenking, B.Sc
Head of Biology department, Maynard School, Exeter
Ann Boyce, B.Sc
Biology department, Maynard School, Exeter

MACMILLAN

First published 1979
Reprinted 1981, 1982, 1983, 1984, 1985

Published by
MACMILLAN EDUCATION LTD
Houndmills, Basingstoke, Hampshire RG21 2XS
and London
Companies and representatives
throughout the world

Printed in Hong Kong

British Library Cataloguing in Publication Data
Jenking, C M
The diversity of life,-(Foundations of biology),
1. Biology
I. Title II. Boyce, A III. Series
574 QH308.7
ISBN 0-333-24193-2

Contents

Preface

Foundations of biology aims to provide a complete pre-university course in biological science. Accordingly, the work is covered in a few handy volumes, not in a single bulky one or numerous monographs. The questions at the ends of the chapters are to test comprehension of the material covered in the chapters and their contents are not necessarily similar to those set in biological examinations which often require knowledge in several branches of biology if they are to be answered properly. Suggestions are provided for further reading.

The course consists of five books written by experienced teachers with special knowledge of biological science, who believe through their experience that fresh approaches to teaching biology are desirable at pre-university level. The books in the series are:

The diversity of life
The cell concept
Heredity, development and evolution
Metabolism, movement and control
Man and the ecosystem.

Biologists will realise the difficulty of subdividing the course into a number of books and opinions will undoubtedly differ on how it should best be done. One difficulty is that a number of topics are based upon knowledge of others, so that if each book is to be helpful some overlap must occur with others in the series. In fact, the necessity for overlap has proved to be relatively small and where it occurs the treatment of topics is consistent from one book to another.

It is wise to remember that no branch of science is more 'fundamental' than any other, so no suggestion has been made that the books need to be studied in a given order. Teachers will be free to use them in any sequence or combination which suits their own courses.

All the authors concerned with the series have felt keenly the inadequacy of purely descriptive biology in giving insight into the basis of science today. It has been necessary therefore for them to introduce some mathematics, physics and organic chemistry to which biology is so closely related. The names of chemical compounds are accompanied by their new names under the IUPAC rules and in *The cell concept* there is an introduction to the new uses which seem difficult at first but which are in fact logical and easy to follow once the principles have been grasped.

The diversity of life includes a wider range of examples of species than is strictly needed by a candidate for any one examination board but because of this the demands of almost all boards are met. This abundance of examples also enables the student to appreciate more fully the characteristics of phyla and the variations within them. The external characteristics of organisms are described in all cases except where knowledge of internal anatomy is required to understand life cycles.

The authors of *The diversity of life* have found in their own teaching a need to append to the study of species concise ecological notes. Accordingly, such notes are included in the book linked with the appropriate illustrations. The authors urge teachers and students to regard the illustrations as a *guide* to the study of the organisms and not a substitute for them. Whenever possible live specimens should be examined.

L. M. J. Kramer
General Editor

1 Organisation and complexity

LEVELS OF ORGANISATION

The purpose of this book is to give some indication of the very wide variety of animals and plants found on the earth today. The book is concerned mainly with their external features and the ways in which they have become adapted to their own particular environments. However, if the relationships of these organisms to each other are to be understood, we have to know something of the internal organisation of their bodies.

Our survey of living things begins with the simplest plants and animals and ends with the most complex ones, the flowering plants and mammals. There is a bewildering assortment of life in between, but it is possible to see certain levels of organisation being attained. These are the cellular, tissue and organ levels of organisation.

The cellular level

The cellular level of organisation is represented by the unicellular animals and plants. Here the entire body consists of a single cell, which must therefore be able to carry out all the functions necessary for life. The term *cell* is normally used as the basic unit of which an animal or plant body is made. It consists of a region of cytoplasm controlled by a single nucleus and enclosed by a cytoplasmic membrane. Plant cells are also surrounded by a box-like *cellulose cell wall*. However, many unicellular organisms, such as the ciliated Protozoa have more than one nucleus controlling their cytoplasm. The term *acellular*, or *non-cellular* is sometimes used for them, meaning that their bodies are not divided up into units or cells.

Some unicellular organisms have a very elaborate internal organisation even though only one unit of cytoplasm is involved, (see *Paramecium* p. 26). This is achieved by having parts of the cytoplasm modified as organelles which perform definite functions. These include the ingestion of food, removal of water and locomotion. These unicellular animals illustrate, in a simplified form the beginning of *division of labour*.

One order of the Algae, the Volvocales, illustrates the development of a colonial organisation (Figure 1a). The order contains unicellular flagellates, such as *Chlamydomonas*, but also a range of increasingly differentiated colonies of flagellates, embedded in mucilage. *Volvox* in particular has some cells of the colony set aside for reproduction and as a result of cytoplasmic connections between the individual cells is able to move in a co-ordinated way. Nevertheless, despite some interdependence of the cells, the colony remains at the cellular level of organisation.

Amongst the animals the sponges are an example of colonial organisation (Figure 1b). Unlike the cells of multicellular animals, those of a sponge remain independent of each other and there is no co-ordinating nervous system. There are five types of cell in a sponge: *epithelial cells* cover the surface; *pore cells* form openings allowing water into the body; *flagellate collar cells* line the internal cavity and aid the flow of water; *mesenchyme cells* are motile and secrete *silicious spicules* which strengthen the colony's wall; *amoebocytes*, also motile, transport food and

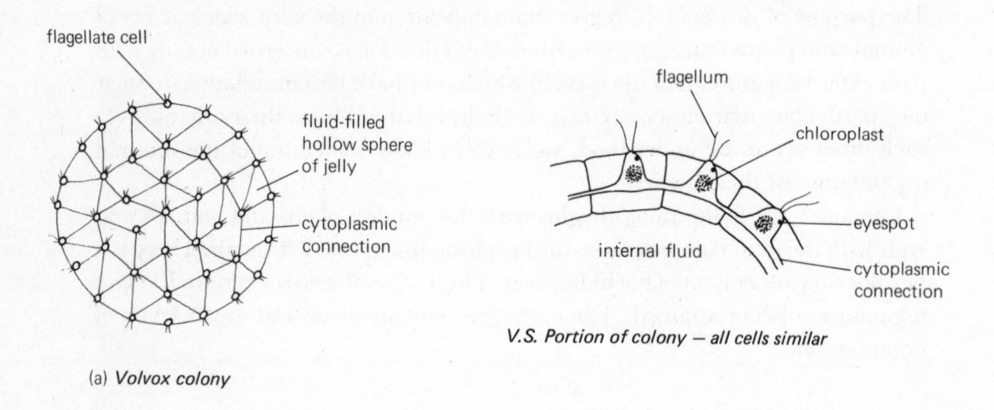

(a) *Volvox colony*

inhalent current enters paragaster (internal cavity)
through pore cells
exhalent current
from paragaster
rock
or
algal
substratum
collar cell from
lining of paragaster
hollow pore cell
allows water into
internal cavity
(paragaster)
spicule-secreting cells
reinforcing the wall
smooth flat
epidermal cells
making up the wall
amoebocyte – motile

(b) Part of a simple sponge colony showing the five types of cell

Figure 1 Colonial organisation

excretory products around the sponge or may become gametes. The flagellate collar cells closely resemble the collar flagellates of the Protozoa and it has been suggested that the sponges could have had their origins in this group. Since the organisation of the sponges is so different from that of other multicellular animals they are often placed in a separate subkingdom, the Parazoa, while the single-celled animals are placed in the Protozoa and the multicellular ones in the Metazoa.

Tissue level of organisation

Within the Metazoa are animals with either two or three body layers. The phylum Coelenterata contains *diploblastic* animals like *Hydra* (see p. 50), which have two cellular layers to the body (as shown in Figure 2). There is the outer *ectoderm* and the inner *endoderm*, with a layer of non-cellular jelly – the *mesogloea*, between them. The ectoderm consists of a sheet of similar cells serving as an external covering, with occasional sensory cells and stinging cells. At intervals are groups of unspecialised *interstitial cells* which can replace used stinging cells or become gametes. The endoderm is a layer of cells specialised for digestion and absorption of food. Running through the mesogloea is a *nerve net* which transmits stimuli received from sensory cells in the ectoderm.

These animals illustrate the tissue level of organisation. By *tissue* we mean a region of the body composed of similar cells all modified for a particular function. In the more highly differentiated animals a number of types of tissue may become associated together making a structure which has a definite purpose. These structures are called *organs*.

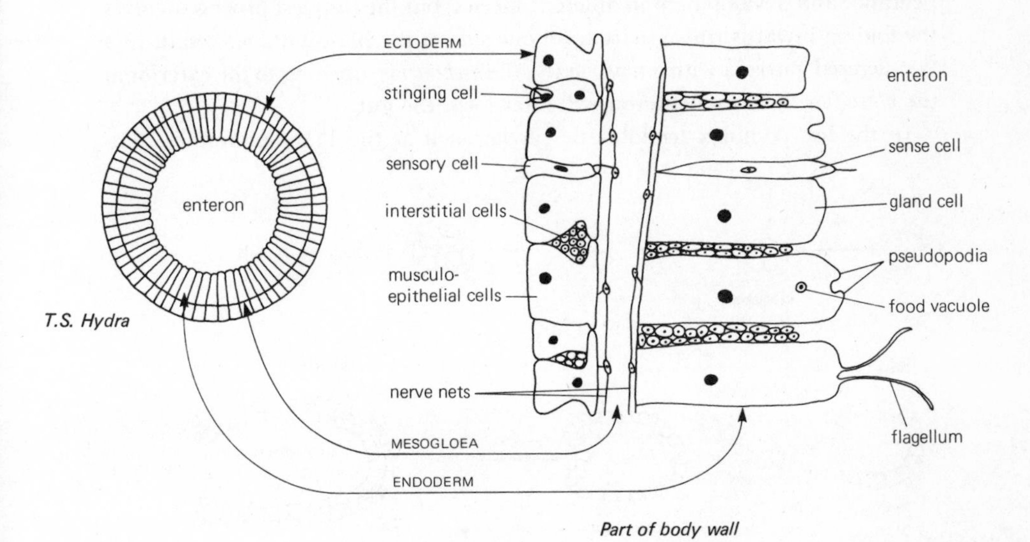

Figure 2 T. S. *Hydra* to show the diploblastic condition and the tissue level of organisation

Most plants are at the tissue level of organisation, having groups of cells arranged as covering, packing, supporting and sometimes conducting tissue. Only the higher plants possess structures such as leaves and flowers, which are organs because they are composed of several types of tissue. The reason why complex organ systems are not required by plants is because of their methods of feeding. Whether photosynthetic, saprophytic or parasitic, the plant does not have to search actively for food or have special parts of the body for ingesting or digesting it.

Organ level of organisation

The development of organs has allowed various parts of an organism to be set aside for a distinct purpose, and division of labour becomes much more obvious. It is associated with the origin of a third body layer, the *mesoderm*, which is present in *triploblastic* animals. This layer is found between the ectoderm and endoderm and may have its beginning in the amoeboid cells found in the mesogloea of some Coelenterata. In the truly triploblastic animals the mesoderm has become the layer which forms the bulk of the body and gives rise to many important organ systems.

Even triploblastic animals go through a stage when they possess only ectoderm and endoderm. As Figure 3 shows the zygote begins its development by undergoing cleavage – dividing by mitosis, into a spherical mass of cells. At first the sphere is solid, but gradually a cavity forms in the centre. At this stage the embryo is a *blastula* and the cavity is the *blastocoel*. The next stage in its development is *gastrulation* – the formation of a *gastrula*. This differs in appearance and development in different species, but the simplest process involves the folding inwards (invagination) of one side of the blastula. This results in a two-layered gastrula with a new cavity, the *archenteron*, opening to the exterior at the *blastopore*. The archenteron will later form the gut.

In the less complex triploblastic phyla, such as the Platyhelminthes (flat-

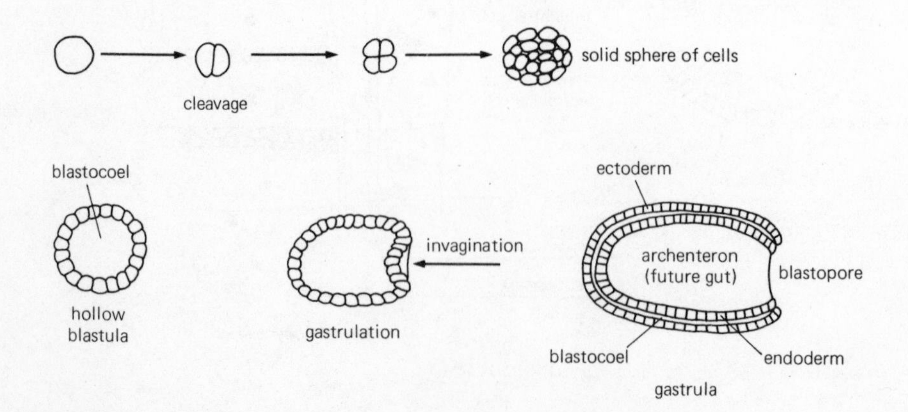

Figure 3 Typical cleavage, blastula and gastrula of a simple chordate

worms, flukes and tapeworms) the mesoderm develops from cells which migrate from the ectoderm and endoderm. In most other phyla some of the mesoderm arises as outgrowths of the wall of the archenteron.

DEVELOPMENT OF THE COELOM

The Platyhelminthes are an example of a triploblastic phylum which is *acoelomate*. This means that its members have mesoderm in which no body cavity or *coelom* develops. Most animal groups do develop a coelom and are therefore *coelomate*. The coelom may arise as pouches of the archenteron – an *enterocoel*; or a cavity may develop within the mass of mesoderm – a *schizocoel* (Figure 4). In most phyla the coelom remains the main body cavity of the adult animal.

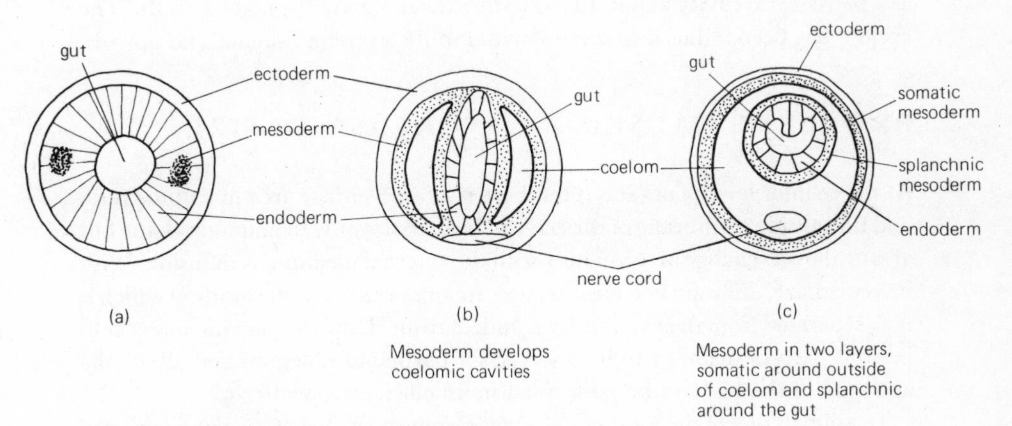

(a) *to* (c) *Development of a schizocoelic coelom in earthworm*

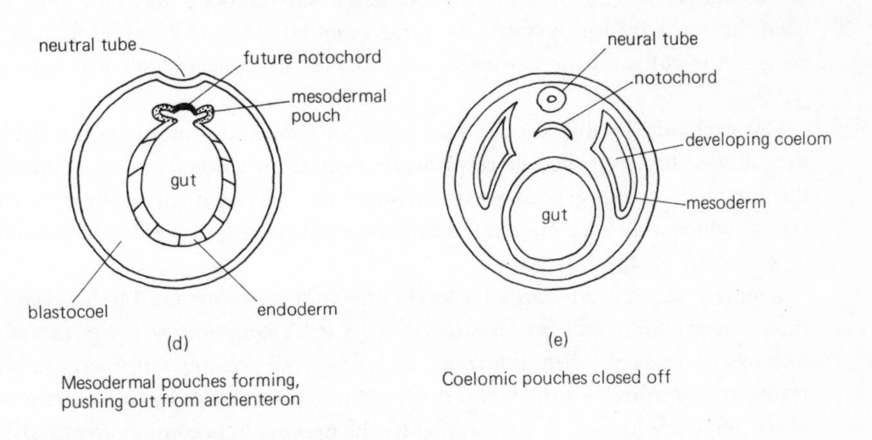

(d), (e) *Development of enterocoelom in a vertebrate*

Figure 4 Methods of coelom development

Possession of a coelom allows the separation of the gut from the body wall, so that the musculature of each can move independently. The coelom contains a *coelomic fluid* which bathes the internal organs and allows them to move slightly in relation to each other without friction. In most phyla the coelom is divided up into one or more cavities around the heart, alimentary canal and other organs.

The haemocoel

The *haemocoel* represents the remains of the blastocoel after the formation of the mesoderm. It usually forms a branching system of tubes – the *vascular system*. In it is liquid containing free mesenchyme cells which give rise to the blood and lymph. Usually one part of the system becomes muscular and acts as a heart, circulating the fluid through the tubes.

In the arthropods, however, the haemocoel has taken the place of the coelom as a perivisceral cavity and its blood bathes the organs of the body directly. The coelom has been reduced to small cavities in the excretory organs and gonads.

PROBLEMS CAUSED BY INCREASE IN SIZE

At the cellular level of organisation there is a large surface area to volume ratio and the outside membrane of the cell can be used not only to maintain shape but also to allow exchange of substances with the external medium by diffusion. With increased size, diffusion becomes too slow to support a body, the inside of which is now separated from the exterior by a bulk of tissue. However, having many cells permits parts of the body to be reserved for particular functions and allows the activities of the body to be performed more efficiently (Figure 5).

In animals one of the most obvious developments is that of an *alimentary canal* which receives and digests food. The simplest, like that of coelenterates has only one entrance/exit, the mouth. Since some parts of the body may be a long way from the gut, a transport system becomes essential. This may be merely projections of the gut as in the Coelenterata and Platyhelminthes, or a *blood vascular system*.

Gas exchange by diffusion also becomes too slow and is improved by having the diffusion limited to special respiratory surfaces such as *gills* or *lungs*. Similarly the removal of nitrogenous waste requires the development of an excretory system which may vary in complexity from the *flame cells* of the Platyhelminthes to the *kidneys* of the Vertebrata.

Animals which must search for food and avoid predators need to be aware of their environment and be able to react to it. A *neuro-sensory system* therefore becomes necessary, with information being received by sense organs and transmitted throughout the body. A simple version is shown by the coelenterate nerve net, but with increasing complexity the nerve cells become aggregated into a *central nervous system*, allowing a definite action instead of a diffuse response. Eventually part of the central nervous system may become a *brain* in overall control of the organism's activities.

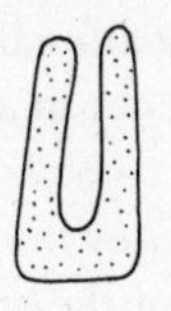

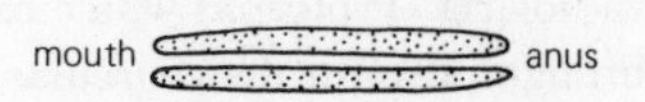

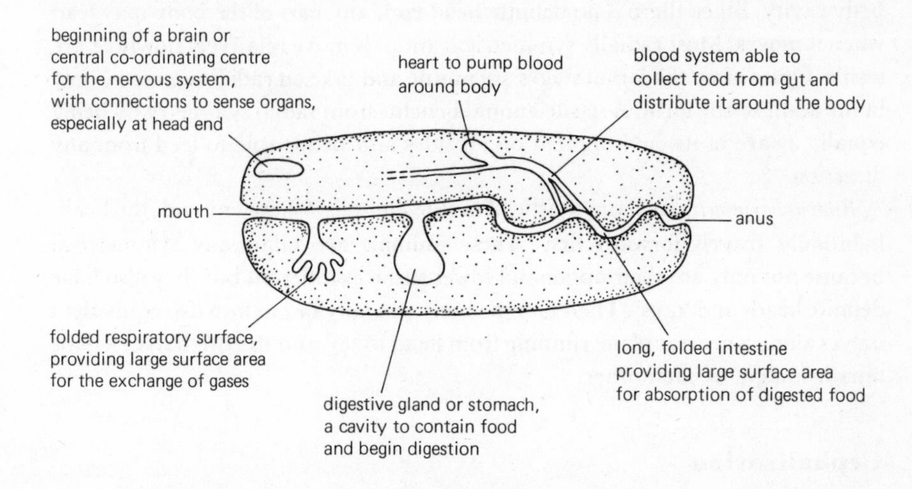

Figure 5 The development of the organ level of organisation

The specialisation of the body (*somatic*) cells in these ways means that the reproductive cells must also be produced in separate organs – the sex organs or *gonads*, which then become definite and permanent structures unlike the transient ones of some Coelenterata.

Some of these problems are shared by plants. That of transport for example, has resulted in the development in the larger plants of vascular tissue, *xylem* for water transport and *phloem* for the carriage of food. Complex organs for feeding, gas exchange and excretion are not required by plants. In green plants, the leaves are organs for both food production and gas exchange. Diffusion is adequate for gas exchange because of the relatively low metabolic rate to be maintained. Feeding in plants, whether by photosynthesis, saprophytism or parasitism results in very little excretory material. Nor is a neuro-sensory system necessary for non-motile plants, which can make only limited responses to their environment, for example by tropic and nastic movements.

SYMMETRY IN ANIMALS

Some Protozoa and most sponges are asymmetrical but most animals show some kind of symmetry. There are three basic types of symmetry: spherical, radial and bilateral. *Spherical symmetry* is shown mainly by free-floating animals such as the Radiolaria (Protozoa) which can feed by pseudopodia put out in any direction through pores in their circular shell.

Radial symmetry is well-illustrated by the Coelenterata. They may be cut into two identical halves by a vertical cut through the centre of the animal. Their radial symmetry is the result of the arrangement of the body around a central body cavity. Since there is no definite head end, any part of the body may lead when it moves. Most radially symmetrical animals move relatively slowly or are sessile. Quite often the larval stages are motile and take on radial symmetry only in the adult sessile form. A sessile animal benefits from radial symmetry by being equally aware of its environment on all sides and being able to feed from any direction.

Bilateral symmetry is characteristic of motile animals, where one end, the head, habitually travels forward first. These animals are bilaterally symmetrical because not only are their upper and lower surfaces different but they also have definite heads and 'tails'. Therefore the body can only be cut into two equivalent halves along a central plane running from head to tail and the two halves will be mirror images of each other.

Cephalisation

Since the head end is likely to come into contact with new stimuli before the rest of the body, there has been a tendency for this part of the body to be well-supplied with sense organs. These provide information about the environment in front of the animal and serve to warn it of danger, presence of food etc. The head may therefore become different in appearance and distinct from the rest of the body. This is called *cephalisation* and varies with the complexity of the animal and the degree of activity it shows.

SEGMENTATION

Metameric segmentation is the repetition of a succession of units or segments along the body of an animal. No animal shows this throughout the whole of its body but the annelids, e.g. earthworms, are a particularly good example. Each segment is very similar to those in front and behind, usually containing portions of body wall, coelom, gut, nerve cord and segmentally-arranged blood vessels, nerves and excretory organs. The segments are separated by partitions called *septa*, but they still function together.

Segmentation begins during the embryonic or larval stage. The number of segments is usually fixed for a particular species and the development proceeds along a set pattern. With the exception of the molluscs, most of the Metazoan phyla show some sign of segmentation during their development. This may be obscured in the adult, however, as in the arthropods and vertebrates where only some organs retain signs of segmentation. Nor does the presence of segmentation indicate relationships between one group of animals and another since it seems likely that the process arose independently several times along different evolutionary lines.

QUESTIONS

1 Distinguish between the following pairs of terms:
   gastrula and blastula;
   enterocoel and schizocoel;
   coelom and haemocoel.
2 Comment on the advantages and disadvantages of a) a third body layer, b) cephalisation.
3 Draw diagrammatic transverse sections of a diploblastic animal, a triploblastic acoelomate, a triploblastic coelomate.
4 To what extent does a green plant exhibit division of labour?

2 Naming and classifying organisms

When first confronted with the task of naming and classifying organisms you may wonder whether such seemingly complex systems are really necessary. In this chapter the principles and aims of classification and naming will be examined with the intention of demonstrating their value.

At present more than one million animal species and nearly half a million plant species have been identified and new species are constantly being discovered. Faced with this diversity man first assigns names to the different kinds of organism and then orders them into groups on the basis of shared characteristics. The latter exercise provides a temporary framework for further studies, around which information about the organisms can be built up. The framework is temporary in that new information may alter the groupings.

Before going any further it will be as well to define a few of the terms that will be encountered. Starting with *classification*, this is the process of ordering organisms into groups on the basis of affinity or relationship between them. *Taxonomy* is the study of the theory, procedures and rules of classification, while *nomenclature* is the naming of the groups used in classification schemes.

NAMING ORGANISMS

Assigning a common name to things of the same kind forms the basis for all human communication. However, biological studies require a particularly precise and universally-applicable system of naming. Comparisons of observations made at different times and in different places are meaningless unless it can be established that the observations refer to exactly the same type of organism. The *binomial system of nomenclature*, introduced by the Swedish scientist, Carolus Linnaeus (1707-78), answered this need. It first appeared in the tenth edition of his important work, *Systema Naturae*, published in 1758. This work also provided the basis for much of the later work in classification.

In the binomial system each species is designated by two latinised names, which together comprise its *scientific* name. The first part of the binomen is termed the *generic* name and is common to the group of closely related species – the genus, to which the organism belongs. The second name denotes the particular kind of organism or species in that genus and is termed the *specific* name. Thus, as can be seen below, the stoat shares its generic name, *Mustela*, with several closely related animals, but is distinguished from these others by its specific name.

	Scientific name	
	Generic name	Specific name
Stoat	*Mustela*	*erminea*
Weasel	*Mustela*	*nivalis*
Pine marten	*Mustela*	*martes*
Polecat	*Mustela*	*putoris*

The generic name is always written with a capital initial letter and the specific name with a small initial letter. The scientific name is always printed in italics and should be underlined in typescript and manuscript. In scientific literature the name of the person who first described the species is often added, so that reference can be made to the original description should there be any difficulty in establishing the identity of an organism.

Scientific names are not generally selected haphazardly. The words chosen often reflect some attribute of the organism. For example, *Homo sapiens* means wise man, *Mytilus edulis* (common mussel) means edible mussel, and *Turdus migratorius* means migratory thrush.

It could be argued that these latinised names are difficult to pronounce and remember, and that nowadays it should be possible to use common or vernacular names and refer to dictionaries for equivalents in other languages. However, this is not possible – the majority of species are not known by a common name. Where common names are used they do not always reflect differences between organisms accurately. For example, birds called robins are found in North America and Great Britain but they are not the same species; indeed the large thrush-like American robin has little in common with the British robin beyond red feathering on the throat and chest.

CLASSIFICATION

Classification has a very long history. Early man must soon have learnt to group animals and plants according to their useful or dangerous properties. However it was not until the seventeenth century that John Ray, a British naturalist, advocated the use of a 'natural' system of classification based on the real natures or natural affinities of organisms, rather than the purely utilitarian or practical systems then in use. Linnaeus had similar aims in mind when he introduced his system of classification based on similarities of form (morphology). Classification based on similarity of form is termed *phenetic*.

Linnaean classification was hierarchical. Organisms were placed into progressively larger and more inclusive groups according to the degree of similarity between them. Thus individuals of the same kind were grouped into *species*, similar species into *genera* (sing. genus), similar genera into *orders* and similar orders into *classes*. The categories or *taxa* (sing. taxon) employed by Linnaeus are still used today, although the discovery of ever-increasing numbers of species has necessitated the introduction of new taxa.

The largest and most inclusive categories are the kingdoms. In most classification schemes the living world is divided into two main kingdoms, the animal kingdom and the plant kingdom. The main taxa in use today are illustrated by reference to the classification of a wallflower (*Cheiranthus cheiri*) and man (*Homo sapiens*) in Table 1.

Table 1

Wallflower		Modern man	
Kingdom	Plantae	Kingdom	Animalia
Phylum	Spermatophyta	Phylum	Chordata
Class	Angiospermae	Sub-phylum	Vertebrata
Order	Rhoeadales	Class	Mammalia
Family	Cruciferae	Infra-class	Eutheria
Genus	*Cheiranthus*	Order	Primates
Species	*cheiri*	Family	Hominidae
		Genus	*Homo*
		Species	*sapiens*

The number of taxa used varies according to the size of the phylum.

It is important to realise that classifications are not rigid systems. Apart from the species none of the categories exists in any real or concrete sense. They are simply groupings or 'pigeon-holes' created by man for his own convenience and as such are continually being revised in the light of new discoveries or fresh interpretations. The major categories – the phylum and the class – are particularly susceptible to change.

As already stated, in Linnaean classification relationship was established on the basis of similar morphology. It was not until the publication of Darwin's theory of evolution that a possible explanation for these patterns of relationship became apparent – namely that organisms are similar because they share a common ancestry and are therefore related by descent. Thereafter the aim of most taxonomists was to construct classifications that reflected evolutionary relationships – such classifications are termed *phylogenetic*. This proved difficult in practice because, although similar morphology suggests a similar origin, the only way to establish this with any certainty is to examine the fossil history of the organisms concerned. Unfortunately, with the exception of the vertebrates, the fossil record of most organisms is incomplete. As a result, although most of the schemes in use today are thought to reflect evolutionary relationships there is often little or no direct evidence for these beliefs.

New disciplines and techniques such as comparative biochemistry, behavioural studies, comparative embryology, comparative serology (study of blood proteins) and cytogenetics have provided a wealth of information which gives a more complete picture of the degree of similarity between organisms. Such a broad-based system is likely to reflect phylogeny far more accurately than one based only on morphology. It will also be a more useful system. Bearing in mind that classifications serve as a means of storing and retrieving information, it follows that the more information used in compiling a classification, the more useful that system will be.

A criticism levelled at many classifications is that the techniques used are too subjective and rely too heavily on the judgements of individual taxonomists. For example, the characters on which a classification is based are usually pre-selected or weighted, that is, certain characteristics are assumed to be better indicators of relationship than others. This is illustrated in the Algae where the classes are separated according to the colour of the dominant pigment, the cell wall material and the nature of the food reserves. The other characteristics of the organisms such as their anatomy and methods of reproduction are largely ignored, on the assumption that they reveal little about relationships within the group.

Recently taxonomists have started to use computers to assess overall similarity, in an effort to standardise procedures and remove as much of the subjective element as possible from classification. As many characters as possible of the organisms being studied are taken into account. The organisms are then compared with reference to all these characters, and the relationship is expressed in terms of percentage similarity. This approach to classification is termed *numerical taxonomy*. Such techniques have proved especially useful in dealing with groups such as bacteria, in which there is very little in the way of morphological features on which to base a classification.

On the following pages the classification used in this book is given in outline only. In the chapters that follow, the main groups of plants and animals are examined in greater detail.

OUTLINE CLASSIFICATION OF THE ANIMAL KINGDOM

Phylum Protozoa Single-celled animals; about 50000 species
All protozoa have a unicellular body. Most are microscopic and all live in moist conditions. Some species contain chlorophyll and so are also classified with the Algae. Protozoa are thought to be the most primitive animals. There are four classes:

Class Flagellata
Class Sarcodina
Class Sporozoa
Class Ciliata

Phylum Porifera Sponges; about 5000 species
Most species are marine, a few live in freshwater. These multicellular organisms probably evolved independently of other multicellular organisms. The body contains a cavity connected to the exterior by pores. Flagellated collar cells line the cavity and create a water current which draws in food particles and oxygen.

There is usually an internal skeleton of calcareous or silicious spicules or horny fibres of spongin. There are three classes:

Class Calcarea (calcareous sponges)
Class Hexactinellida (glass sponges)
Class Demospongiae (horny sponges)

Phylum Coelenterata Hydroids, jellyfish, corals and sea anenomes; 9400 species
These aquatic species are all radially symmetrical. The body is made up of two layers of cells (diploblastic) surrounding a cavity–the enteron. A single opening, the mouth, is surrounded by tentacles which bear stinging cells used in food capture. There are two main structural types, sessile hydroids or polyps, and umbrella-shaped, free-swimming medusae. Some forms pass through both stages in their life cycle. Polyps typically reproduce asexually by budding. Medusae typically reproduce sexually, giving rise to a characteristic planula larva. There are three classes:

Class Hydrozoa (hydroids)
Class Scyphozoa (jellyfish)
Class Anthozoa (sea anenomes and corals)

From this point onwards all the groups are *triploblastic*, that is, the body is made up of three layers of cells.

Phylum Platyhelminthes Flatworms, flukes and tapeworms; about 5500 species
These bilaterally symmetrical, flattened, worm-like animals lack a body cavity. Some are free-living, many are parasitic. All possess flame cells for excretion and osmoregulation. There are three classes:

Class Turbellaria (free-living flatworms)
Class Trematoda (flukes)
Class Cestoda (tapeworms)

Phylum Nematoda Roundworms; about 10000 species
Some nematodes are free-living, others are parasitic. The body is bilaterally symmetrical and elongate, tapering at each end. The body covering is a thick elastic protein cuticle. The sexes are usually separate.

From this point onwards all the groups possess a true body cavity or *coelom*.

Phylum Annelida Ringed or segmented worms; nearly 7000 species
The body is divided into similar segments (metamerism), separated internally by sheet-like septa. In most species each segment bears bristle-like chaetae. There is a well-developed body cavity or coelom. Nephridia are typically the excretory and osmoregulatory organs. There are three classes:

Class Polychaeta (bristle worms)
Class Oligochaeta (earthworms)
Class Hirudinea (leeches)

Phylum Mollusca Molluscs; about 45000 species
The body is unsegmented and divided into the following regions: a head, a ventral muscular foot and a dorsal visceral hump, which contains most of the organs. The visceral hump is generally protected by a calcareous shell secreted by the mantle (a layer of skin which covers the hump and hangs down over the foot). Many molluscs have a rasping, tongue-like radula for feeding. Fertilisation typically results in the formation of a trochophore larva (as in polychaetes). There are six classes:

Class Monoplacophora
Class Amphineura (chitons)
Class Gastropoda (slugs, snails and limpets)
Class Scaphopoda (tusk shells)
Class Lamellibranchia (bivalves)
Class Cephalopoda (cuttlefish, squids and octopuses)

Phylum Arthropoda Arthropods; about 750 000 species
Arthropods have segmented bodies encased in a horny cuticle which forms an exoskeleton. Many of the segments bear paired, jointed appendages, e.g. legs and mouthparts. There is a head which bears the mouthparts and sense organs. There are six main classes:

Class Crustacea (shrimps, crabs, lobsters etc.)
Class Merostomata (king crabs)
Class Arachnida (spiders, scorpions, ticks etc.)
Class Onycophora (velvet worms)
Class Myriapoda (centipedes and millipedes)
Class Insecta (insects)

Phylum Echinodermata Starfish, sea urchins, sea lilies, brittle stars and sea cucumbers; about 5400 species
All echinoderms are marine. They have no head and no true brain. Typically adults display a five-rayed radial symmetry. The skin contains chalky plates and sometimes spines. The mouth is usually on the ventral surface of the body. There are five classes:

Class Asteroidea (starfish)
Class Ophiuroidea (brittle stars)
Class Echinoidea (sea urchins)
Class Holothuroidea (sea cucumbers)
Class Crinoidea (sea lilies)

Phylum Chordata Chordates; about 51 000 species
At some time during their life history all chordates possess the following features:

1 a dorsally situated, strengthening rod – the notochord;
2 a hollow dorsal nerve cord;
3 pharyngeal clefts;
4 a closed blood vascular system;
5 a post anal tail.

This phylum includes three sub-phyla, two of which, the Cephalochordata and the Tunicata, lack a backbone. In the third sub-phylum, the Vertebrata, the notochord is replaced by a vertebral column. Vertebrates are sub-divided into seven classes:

Class Agnatha (jawless fish)
Class Chondrichthyes (cartilaginous fish)
Class Osteichthyes (bony fish)
Class Amphibia (amphibians)
Class Reptilia (reptiles)
Class Aves (birds)
Class Mammalia (mammals)

OUTLINE CLASSIFICATION OF THE PLANT KINGDOM

Phylum Bacteria About 1200 species
Bacteria are almost ubiquitous in their occurrence. They are all extremely small, unicellular organisms. The diameter of bacterial cells is generally about 1 μm. They have a very simple structure. There are no mitochondria, chloroplasts or nuclear membranes. The bacterial cell is surrounded by a cell wall which, in some species, is enclosed in a capsule. Some bacteria bear flagella, fine thread-like locomotory organelles. Nutrition is varied. Some are autotrophs (self-feeding) and manufacture their own food using the energy from sunlight (photosynthesis) or the energy released from chemicals in their environment (chemosynthesis). The rest are mainly parasites or saprophytes. The former cause many human diseases such as tuberculosis, pneumonia, cholera and plague. Saprophytic bacteria play an extremely important part in recycling nutrients in the soil; thus helping to maintain the soil's fertility. Bacteria reproduce asexually by binary fission. There are several different ways of classifying bacteria. The simplest is according to shape – there are four basic types:

Cocci – spherical cells
Bacilli – cylindrical, rod-shaped cells
Vibrios – comma-shaped rods
Spirilla – corkscrew-shaped rods

As indicated earlier in this chapter, the most recent classification of bacteria is a numerical one, based on all possible characters including biochemistry, physiology and host/parasite relationships where they exist. (Certain aspects of bacterial biology are described in *The cell concept*.)

Phylum Algae About 20000 species
Most algae are aquatic. They all contain photosynthetic pigments. There is no differentiation into root, stem and leaf. Unicells, colonies, filaments and thalloid (sheet-like) forms are all represented. Sex organs are usually unicellular. Asexual reproduction is by binary fission, fragmentation and spores – most commonly

motile zoospores. There are five main classes:

Class Chlorophyceae (green algae)
Class Cyanophyceae (blue-green algae)
Class Bacillariophyceae (diatoms)
Class Rhodophyceae (red algae)
Class Phaeophyceae (brown algae)

Phylum Fungi Moulds, mushrooms and toadstools; about 80000 species
Photosynthetic pigments are absent in all fungi. The majority are either saprophytes or parasites. Most fungi are terrestrial. The plant body consists of a network of threads (hyphae) forming a mycelium. Most have a cell wall containing chitin. Reproduction is by spores which are produced both sexually and asexually. There are three classes:

Class Phycomycetes
Class Ascomycetes
Class Basidiomycetes

Phylum Bryophyta Mosses and liverworts; about 25000 species
The plant body is differentiated into simple leaves and root-like structures, but there is little tissue differentiation. There is well-defined alternation of generation between a transitory, diploid, spore-producing sporophyte and a persistent, haploid, gamete-producing gametophyte. They are generally terrestrial but few can tolerate dry conditions. Male gametes are motile and fertilisation is dependent on the presence of water. There are two classes:

Class Hepaticae (liverworts)
Class Musci (mosses)

Phylum Pteridophyta Ferns, horsetails and club mosses; about 10000 species
Pteridophytes display alternation of generations. The sporophyte is the dominant phase and is differentiated into root, stem and leaves. They exhibit a high degree of tissue differentiation. The sporophyte is well adapted to terrestrial conditions but male gametes are motile and fertilisation is therefore dependent on moist conditions. There are three classes:

Class Filicales (ferns)
Class Equisitales (horsetails)
Class Lycopodiales (club mosses)

Phylum Spermatophyta Seed plants; about 300000 species
The dominant plant is the sporophyte. Spermatophytes display the highest degree of tissue differentiation in the plant kingdom. Fertilisation is by means of a pollen tube (siphonogamous) and is independent of external water. Reproductive organs are borne in special structures, cones or flowers. Fertilisation results in the formation of a protected embryo plant – the seed. There are two classes:

Class Gymnospermae (cone-bearing plants)
Class Angiospermae (flowering plants)

QUESTIONS

1 Why is the binomial system of nomenclature used in scientific literature?

2 Explain briefly what is meant when a system of classification is described as: a) natural; b) phylogenetic; c) numerical.

3 Phylum Protozoa

This phylum contains more than 50000 very diverse species, all of which have unicellular bodies. Although most species live as solitary individuals, there are many colonial forms. Protozoans occur in almost all moist habitats – in the sea, in freshwater, in the soil, and in the bodies of other animals.

'Protozoa' means 'first animals' and members of this phylum are thought to be the most primitive animals, that is, most closely related to the ancestral organism from which plants and animals evolved. This does not mean they are all very simple animals. Many protozoa are highly specialised with their cytoplasm differentiated into organelles of many different kinds, performing a variety of functions.

The unicellular condition imposes a restriction on size; most Protozoa are microscopic. This is partly because they rely on diffusion over the body surface as the means of transporting substances into and out of their bodies, and such a process is only efficient in a small organism where the surface area to volume ratio is sufficiently large (this ratio decreases with increasing size). Protozoans are also restricted in their choice of habitat, being limited to moist situations. The thin moist plasma membrane, essential for diffusion, is obviously unsuited to dry conditions.

Nutrition in this group is very varied. Some Protozoa are *autotrophs* (self-feeding); they contain chlorophyll and, like plants, manufacture their own food from inorganic substances by photosynthesis. However, the majority of species are *heterotrophs*, requiring a supply of ready-made food as they are unable to synthesise their own.

Binary fission is the most common form of asexual reproduction. This is accompanied by sexual processes in many species.

CLASSIFICATION

Protozoans are divided into four classes as follows:

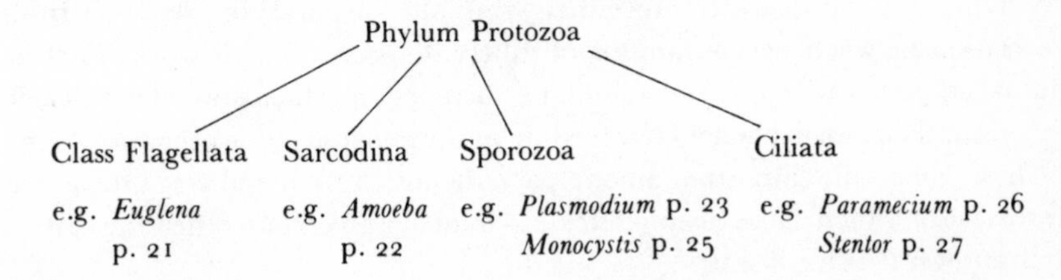

Class Flagellata

Flagellates are thought to be the most primitive protozoans. They all possess one or more flagella for the greater part of their life history. The flagella are the main locomotory organelles. In most flagellates there is a pellicle consisting of protein bands underlying the plasma membrane.

Flagellates are divided into two sub-classes, the Phytoflagellata and the Zooflagellata.

Sub-class Phytoflagellata

This group contains all the forms that are difficult to classify because they possess both plant and animal characteristics. In most classification schemes they are accordingly placed with both the Algae and the Protozoa.

All phytoflagellates are motile, a characteristic usually associated with animals. Some forms lack the cellulose cell wall typical of plants and yet, like plants, most phytoflagellates possess chloroplasts. These organelles contain chlorophyll and varied amounts of xanthophylls. Where chlorophyll predominates, the organisms appear green, where xanthophylls predominate, they appear red, orange, yellow or brown. Although all these coloured forms are capable of photosynthesis, not all are strictly autotrophic, that is, able to build up food molecules from simple inorganic molecules; many can only synthesise food from fairly complex organic molecules. Many species also differ from plants in being unable to synthesise all their own vitamins.

The 'intermediate' status of many members of this group is perhaps most clearly shown by members of the genus *Euglena* (Figure 6).

The dinoflagellates are another interesting group of phytoflagellates. They are often very abundant in marine and freshwater *plankton* (the drifting assemblage of microscopic plants and animals that live suspended near the surface of bodies of water). In certain conditions the population of some marine dinoflagellates increases dramatically (as many as 40 million per cubic centimetre have been recorded), turning the water red, or in some cases orange or brown. The toxic wastes produced by these 'red tides' can cause large scale destruction of marine life.

Sub-class Zooflagellata

These are the animal-like flagellates. They do not possess chloroplasts. Some are free-living but the majority are either symbiotic or parasitic. *Symbiosis* is an association between two organisms of different species, which works to their mutual advantage. *Parasitism* is a similar association, in which however, only one organism, the parasite, benefits by feeding at the expense of the other organism, the host. Especially important among parasitic forms are members of the genus *Trypanosoma*, which cause sleeping sickness in man and certain other animals, as well as several other diseases.

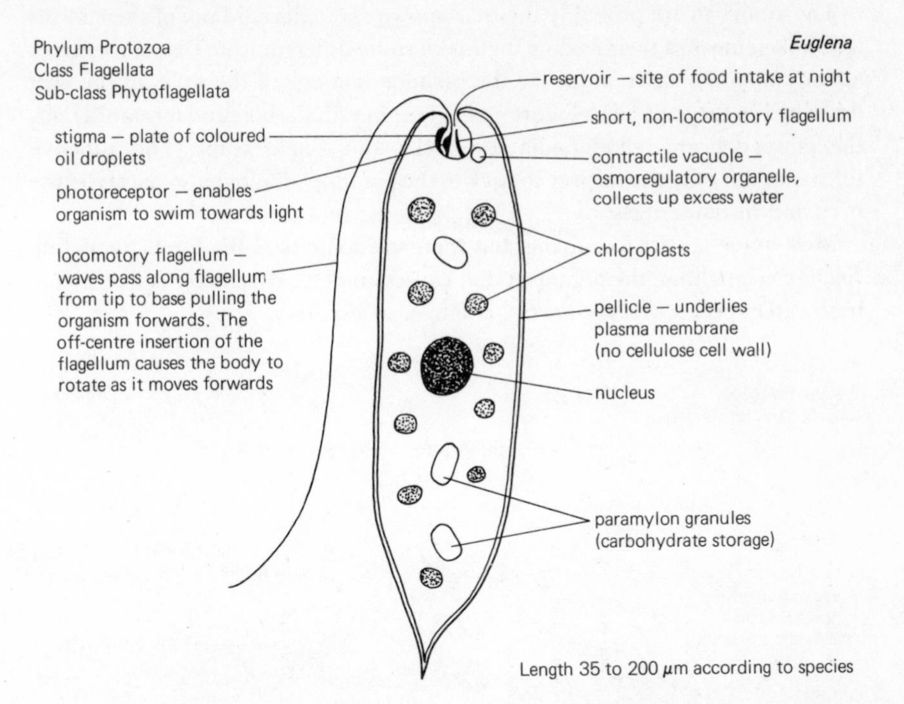

Figure 6 *Euglena*

Species of *Euglena* are very common and widespread. They are especially abundant in ponds and ditches that are rich in organic matter, where they frequently impart a green colour to the water.

Euglena generally moves using the flagellum but it can also change the shape of its body and thus creep over surfaces (euglenoid movement).

In the light *Euglena* is a photosynthetic autotroph; in the dark it absorbs nutrients through the surface membrane in the reservoir. They also differ from most plants in being unable to synthesise vitamin B_{12} and so require an external supply of this vitamin.

Reproduction is by means of binary fission. In unfavourable conditions *Euglena* can round off and encyst.

Class Sarcodina

Sarcodines all possess flowing extensions of the body called *pseudopodia*. These serve as locomotory organelles and also for food capture. The majority of species have a calcareous or silicious skeleton that is either internal, as in radiolarians, or external, as in foramniferans. Radiolarians and foramniferans are both important constituents of marine plankton. Indeed, their abundance is such that their skeletons form the main constituent of many ocean bed sediments. More than 30 million square miles of ocean floor are covered by sediments rich in radiolarian skeletons (radiolarian ooze) alone.

The amoebae are probably the best-known sarcodines. Many of these forms lack a skeleton and their feeding method is quite different from that of the forms already mentioned. In amoebae the pseudopodia engulf the prey, in the way described in the caption to Figure 7, whereas in radiolarians and foramniferans, the more delicate pseudopodia are sticky and act as traps. The adhesive substance not only causes prey to stick to the pseudopodia, but also paralyses the prey and initiates digestion.

Most amoebae are free-living, but there are some parasitic forms, including *Entamoeba histolytica*, the organism that causes amoebic dysentery. A free-living, freshwater species, *Amoeba proteus*, is shown in Figure 7.

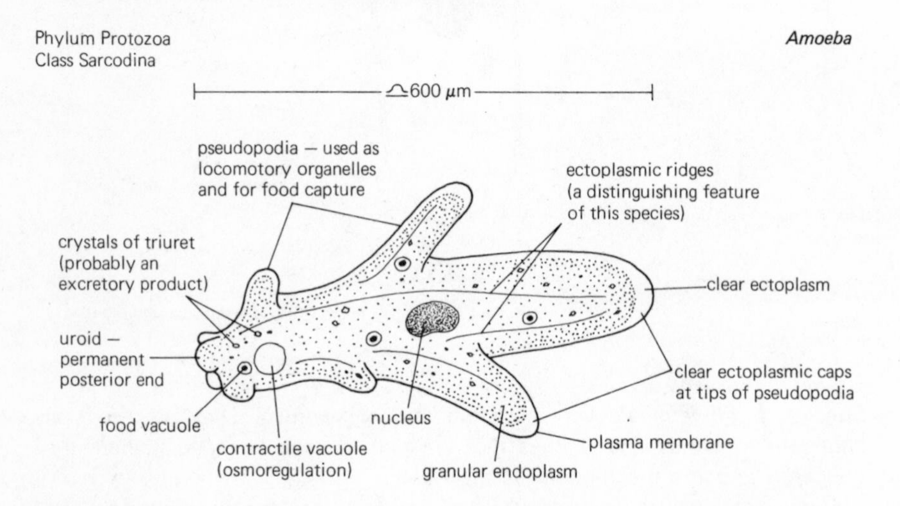

Figure 7 *Amoeba*
Amoeba proteus is found on the bottom of shallow ponds and slow-flowing streams, especially where there is abundant decaying vegetation. Food particles, animal or plant, alive or dead, are ingested by phagocytosis. Two pseudopodia flow out and surround the particle, eventually fusing around it, forming a food vacuole in which digestion occurs. The soluble products of digestion diffuse into the cytoplasm and the organism flows away from any indigestible remains. *Amoeba* reproduces by binary fission. In unfavourable conditions the organism rounds off and secretes a resistant cyst around itself, from which it emerges when favourable conditions return.

Class Sporozoa

These protozoans differ from members of the other classes in that they have no immediately apparent locomotory organelles. All are internal parasites of other animals and have complex life histories, involving both sexual and asexual generations, and sometimes two hosts. After sexual reproduction, the zygote developes into a spore-like structure – hence the name of the group.

Phylum Protozoa
Class Sporozoa

Plasmodium – the malarial parasite

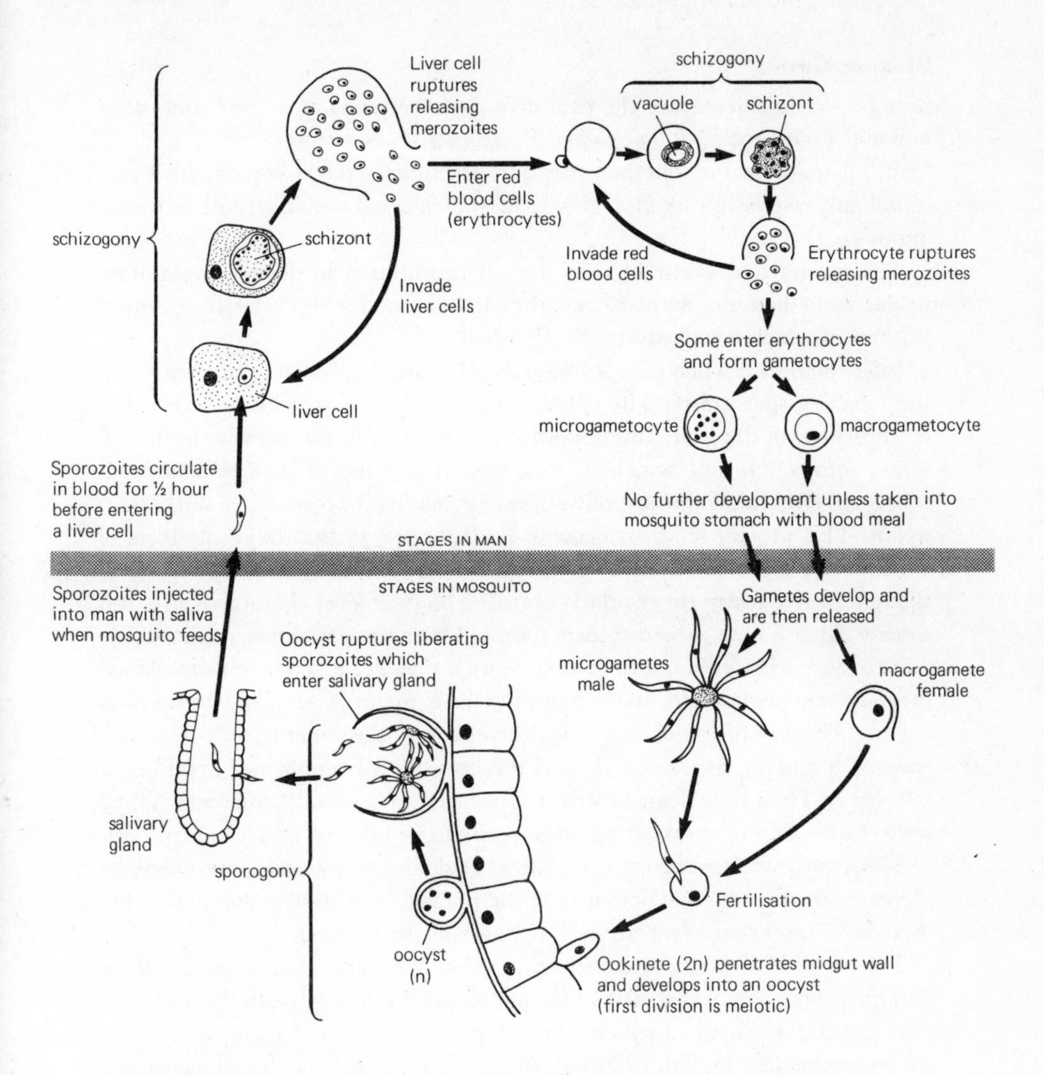

Figure 8 Life cycle of *Plasmodium* (see text for description of life cycle)

Malaria occurs in most countries of the world between latitudes 60° N and 30° S, although tropical Africa is probably the worst affected area. The distribution of malaria is dependent on the distribution of anopheline mosquitoes.

There are two lines of attack used in attempts to control malaria. One is to try to destroy parasites in infected individuals by the use of drugs like quinine. Drugs can effect a complete cure in many cases.

The other method is to destroy the mosquitoes. There are various ways of doing this, such as draining wet areas where mosquitoes breed, or spraying house walls with insecticides. The latter method is probably the most generally useful, although in some areas mosquitoes have developed resistance to many of the insecticides used.

Many sporozoans are important parasites of man and domestic animals. These include *Eimeria*, an important pathogen of cattle and poultry, and *Plasmodium*, the malarial parasite.

Plasmodium

Members of this genus are the causative agents of malaria in man and other animals. Four species, *P. falciparum, P. malariae, P. ovale* and *P. vivax*, occur in man. These species differ in their effects. *P. falciparum* is the most pathogenic and is probably responsible for more deaths in tropical regions than any other disease organism.

The species that occur in man are all transmitted by female mosquitoes belonging to the genus *Anopheles*, and they all have similar life cycles, the essential features of which are illustrated in Figure 8.

Infection occurs when man is bitten by a female *Anopheles* mosquito carrying the infective sporozoites. The sporozoites enter the liver cells, and then the erythrocytes, of the host. Inside both types of host cell the parasite feeds and grows into a schizont which divides into merozoites; a process known as *schizogony*. Eventually the host cells rupture, releasing the merozoites and certain toxins. This process tends to become synchronised so that large numbers of erythrocytes burst simultaneously, releasing large quantities of toxin; this is thought to give rise to the regularly occuring bouts of fever characteristic of this disease. After a time gametocytes are formed from some of the merozoites. The gametocytes will develop into gametes only if they are taken in with the blood meal of a female *Anopheles*. In the mosquito the gametes develop and fertilisation occurs. The resulting ookinete migrates through the midgut wall and then encysts, becoming an oocyst. Inside the oocyst wall sporozoites are formed (*sporogony*). These are released by the rupture of the oocyst wall, after which they make their way to the salivary glands, ready to be injected into a human host.

Schizogony and sporogony are phases of asexual reproduction that serve to increase the reproductive potential of the parasite and so help counteract the inevitable losses that occur at various stages of the life cycle.

The life cycle of another parasitic form, *Monocystis agilis*, is shown in Figure 9. A large proportion of the earthworm population is infected with this parasite, which lives in the seminal vesicles, but its presence seems to have little effect on sperm production, probably because the sperm are produced in such enormous numbers.

Class Ciliata

As the name of the group suggests, these protozoa possess cilia for at least part of their life history. The cilia function as locomotory and food-acquiring organelles. The arrangement of the cilia varies, and on the basis of this the class is divided into two sub-classes, the Holotricha, in which the covering of cilia is uniform, e.g. *Paramecium* (Figure 10), and the Spirotricha, in which it is not, e.g. *Stentor* (Figure 11).

Ciliates typically possess two types of nuclei, macronuclei and micronuclei.

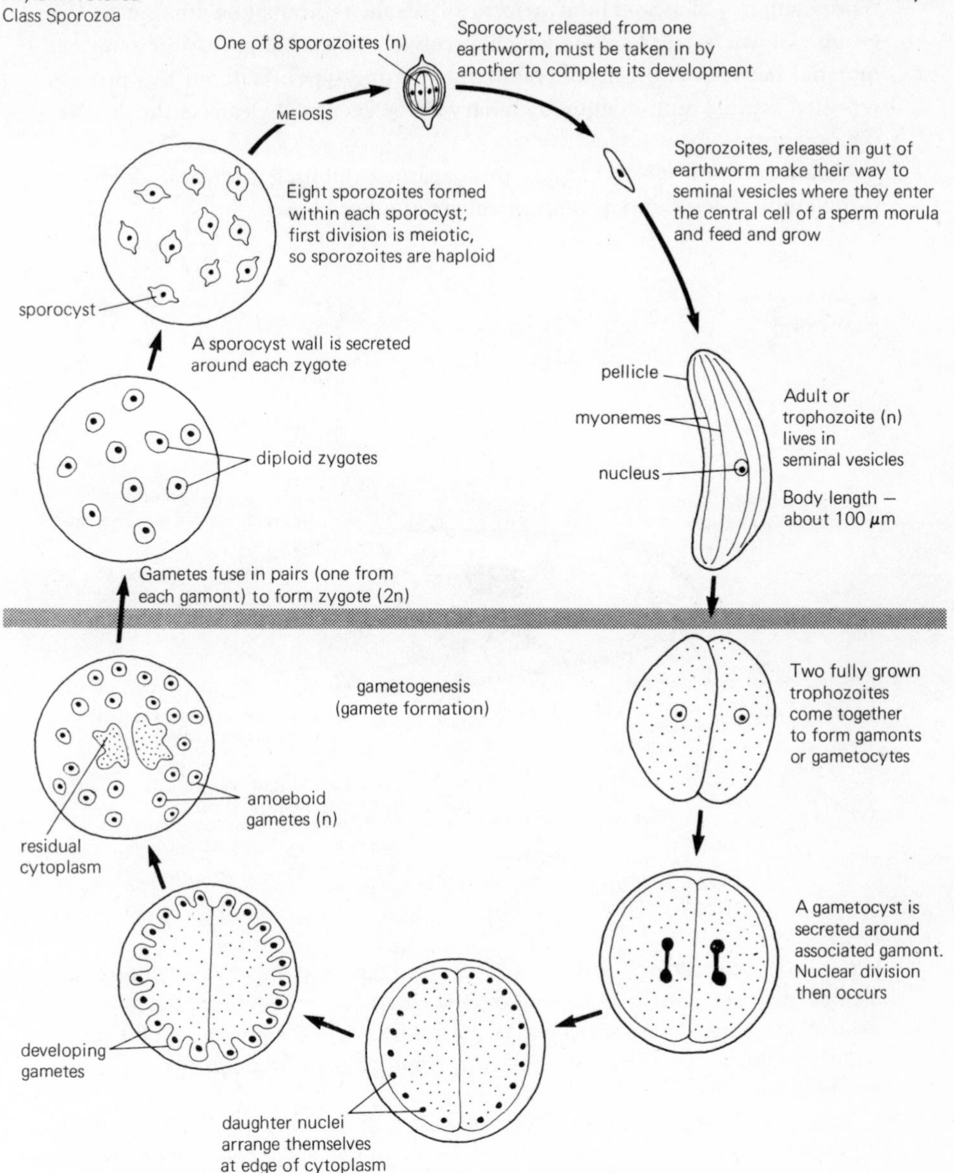

Figure 9 Life cycle of *Monocystis*

In Britain there are more than twenty species of *Monocystis* parasitic in earthworms. Perhaps the best known is *Monocystis agilis* which lives in the seminal vesicles; the life cycle of this species is illustrated above. It is not clear how sporocysts find their way into the soil, other than at the death of the host, although it has been suggested that they may be transferred from one host to another with the sperm during copulation.

The larger *macronucleus* is essential for normal development and functioning of the cell, while the *micronucleus* is responsible for genetic exchange during sexual reproduction. Ciliates exhibit a form of sexual reproduction unique to this group, known as *conjugation*, which involves the exchange of micronuclear material between two cells of compatible mating types. Without this process, repeated asexual reproduction by binary fission eventually leads to the death of the organism.

Ciliates are the most complex protozoans, exhibiting a greater degree of cytoplasmic differentiation than members of other classes.

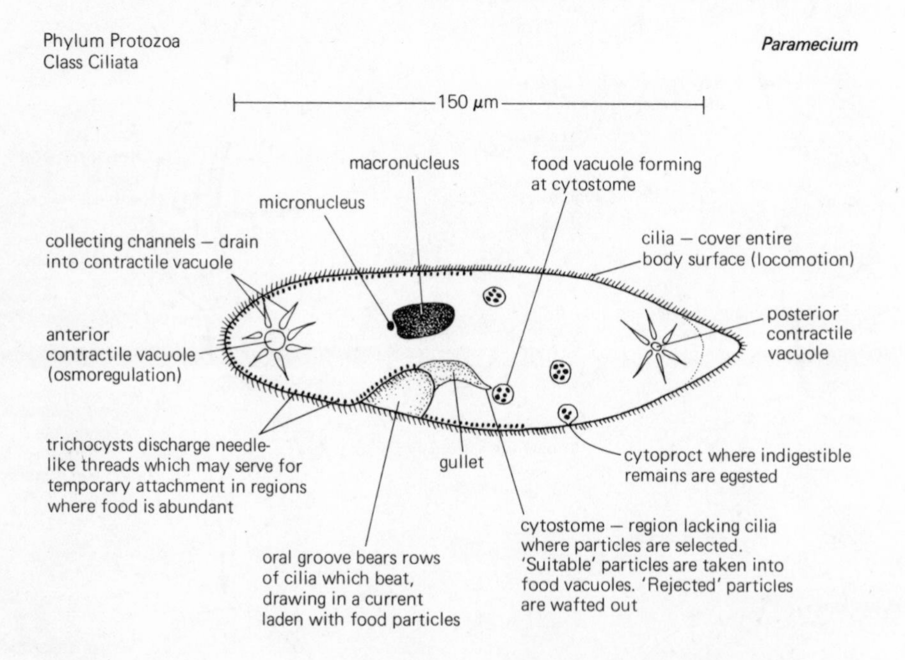

Figure 10 *Paramecium caudatum* is very common in freshwater ponds and ditches that are rich in decaying organic material. It feeds on bacteria and microscopic algae. These are drawn to the cytostome in the way described above, and there taken into food vacuoles. The latter circulate through the cytoplasm where digestion occurs.

Coordinated waves of ciliary movement pass over the body causing the animal to move forwards and also to rotate. The direction of the ciliary beat can be reversed resulting in backwards motion. There is a network of microfibrils (kinetodesmata) connecting the ciliary bases and it is assumed that these help synchronise ciliary beat.

Conjugation usually occurs when food is scarce, providing compatible mating types are present. The oral surfaces of two individuals become confluent (they meet and adjoining membranes break down). They remain like this for 16 to 18 hours during which a complex sequence of events takes place, the essential feature of which is the exchange of micronuclear material. At the end of the process the conjugants part and immediately undergo binary fission.

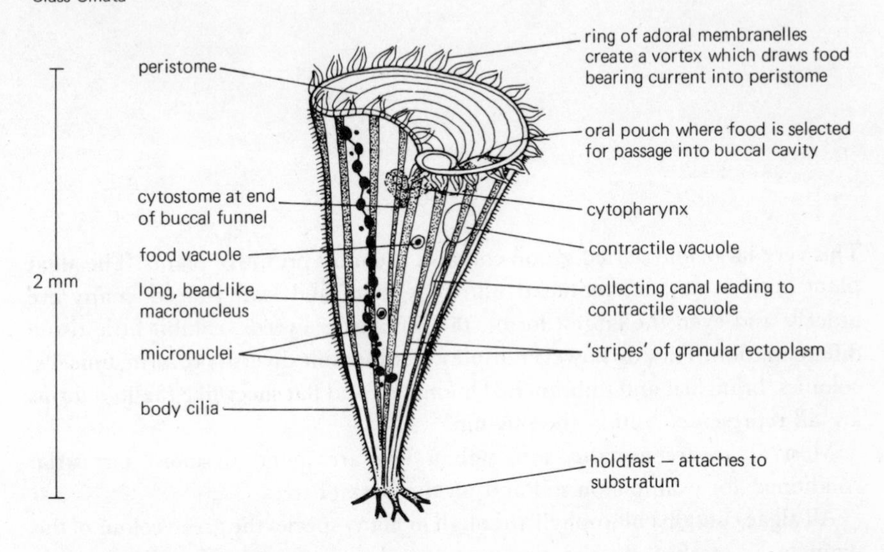

Figure 11 *Stentor coeruleus* is one of the largest known protozoans. It is fairly common in slow-moving freshwater where it normally lives attached to the surface of vegetation, although it is highly contractile and can swim freely. *Stentor* feeds on bacteria and microscopic algae which are drawn into the buccal funnel by the beating of the adoral zone of membranelles. Each membranelle consists of two to three rows of fused cilia.

QUESTIONS

1. Explain why *Euglena* is sometimes described as an 'intermediate' organism.
2. Describe the criteria used to distinguish between the four protozoan classes.
3. Explain why *Paramecium* is said to be a more advanced organism than *Amoeba*.
4. Briefly describe the life cycle of *Plasmodium* and explain how this organism is adapted to its parasitic way of life. You will need to refer to chapter 7.

4 Phylum Algae

This very large and varied group contains the most primitive plants. The algal plant body is not differentiated into root, stem and leaf. Indeed, many are unicells and even the largest forms, the familiar seaweeds, exhibit little tissue differentiation. They do however display considerable diversity of form, unicells, colonies, branched and unbranched filaments, and flat sheet-like thalloid forms are all represented within the phylum.

Most algae are aquatic, although a few are found in moist terrestrial conditions, for example on soil and on the bark of trees.

All algae contain chlorophyll although in many species the green colour of this pigment is masked by the presence of other pigments. Unicellular algae constitute the basis of most food chains in the sea, comprising as they do the autotrophic component of plankton. Indeed the abundance of these organisms is such that they play a vital part in maintaining the balance of gases in the atmosphere. It has been estimated that planktonic organisms are responsible for as much as 70 per cent of all photosynthesis on earth.

Asexual reproduction is usually by fission in unicells, by fragmentation in filamentous forms, or by the production of spores. The latter are either motile, in which case they are termed *zoospores*, or non-motile. Sexual reproduction is varied, being *isogamous* – where the gametes are indistinguishable both in appearance and motility, e.g. *Chlamydomonas*, or *anisogamous* – where the gametes differ slightly in size and/or motility, e.g. *Spirogyra*, or *oogamous* – where the differences between the gametes are very pronounced; the female gamete is immotile and much larger than the motile and more numerous male gametes, e.g. *Fucus*. Life cycles are equally variable, haplont and diplont life cycles, as well as alternation of generation, are all encountered (these terms are explained in chapter 19).

CLASSIFICATION

Algae are classified according to the colour of their dominant pigment, the nature of the cell wall material, and the food storage products utilised by the group. Most classification schemes recognise seven or eight classes of algae, the five detailed on the next page are the most important. It is now generally recognised that many of these classes are not, in fact, very closely related. This has led some authorities to raise each of the classes to the rank of phylum.

Many algae live in symbiotic association with fungi as lichens.

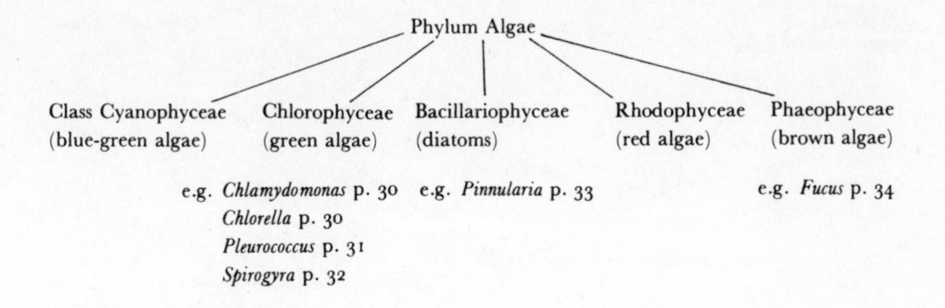

Class Cyanophyceae

This is thought to be one of the most ancient groups. There is some evidence that blue-green algae, very similar to present-day forms, were already in existence three thousand million years ago.

Unicellular, colonial and filamentous forms occur in seawater and freshwater habitats. They all contain the pigments phycoerythrin and phycocyanin and they differ from other algae in having a very simple structure, much like that of bacteria, lacking a membrane-bound nucleus, chloroplasts, mitochondria, Golgi bodies and endoplasmic reticulum.

Class Chlorophyceae

The dominant pigment in this class, as its name suggests, is chlorophyll. The cell wall is made of cellulose, and the main storage material is starch. The most primitive species have flagella, and on the basis of their motility, which is normally considered an animal feature, they are also classified with the protozoan phytoflagellates.

Green algae range in form from simple unicells to quite complex thalloid forms, e.g. *Ulva lactuca* – the sea lettuce, a common species on British shores. The structures of a motile unicell – *Chlamydomonas*, an aquatic non-motile unicell – *Chlorella*, a terrestrial non-motile unicell – *Pleurococcus*, and a filamentous form – *Spirogyra*, are illustrated in Figures 12, 13, 14 and 15.

Chlamydomonas

Approximately twenty species of *Chlamydomonas* are found in Britain. They reproduce asexually by fission. The flagella are withdrawn and the organism divides into four or more daughter cells within the thin, flexible cell wall (cell envelope) of the parent. The daughter cells develop flagella and their own cell envelopes and then burst free from the parent envelope. In certain circumstances the release of the daughter cells is delayed and they go on dividing inside the parent envelope, which becomes increasingly mucilaginous; this 'palmella' stage is thought to be a response to unfavourable conditions.

Sexual reproduction occurs when individuals of different mating types come into contact. The 'adult' or vegetative organisms shed the cellulose envelope to become gametes (either isogametes or anisogametes according to the species),

Phylum Algae
Class Chlorophyceae

Chlamydomonas

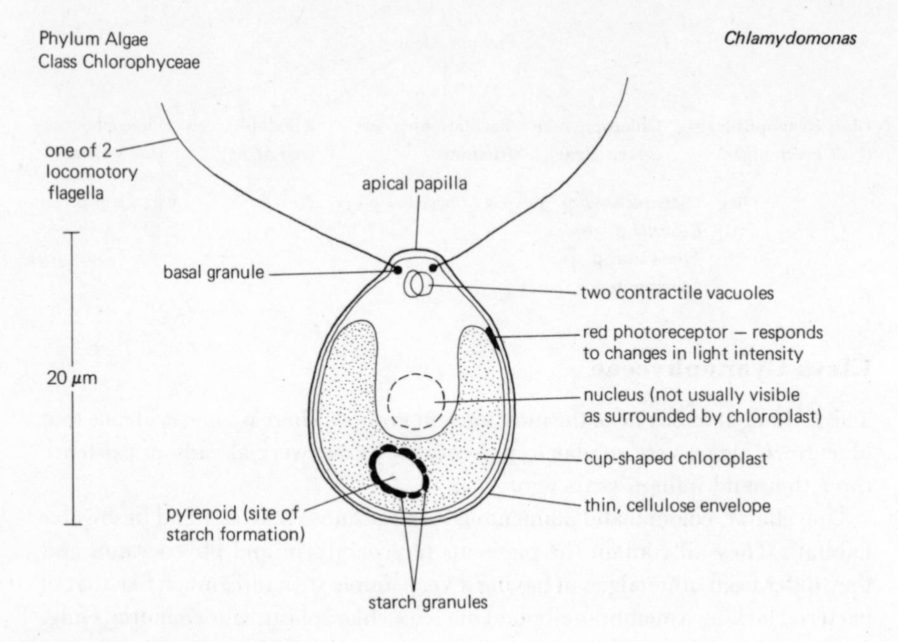

Figure 12 *Chlamydomonas*
Most species live in stagnant, freshwater ponds and ditches, often in such large numbers that they impart a green colour to the water. A few species live in the sea and brackish water.

All species seem to be photosynthetic autotrophs, although there is evidence that some take in organic substances from the environment.

Phylum Algae
Class Chlorophyceae

Chlorella

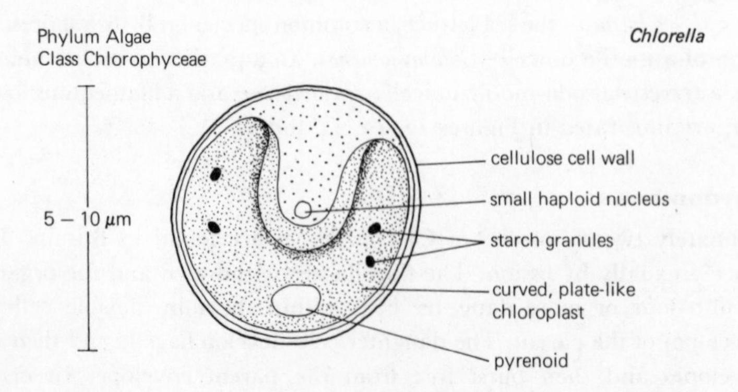

Figure 13 *Chlorella* often abounds in stagnant ponds and ditches.
It is also found living symbiotically in the bodies of certain animals, for example, *Stentor* and *Hydra*. Reproduction is by cell division; there is no sexual reproduction.

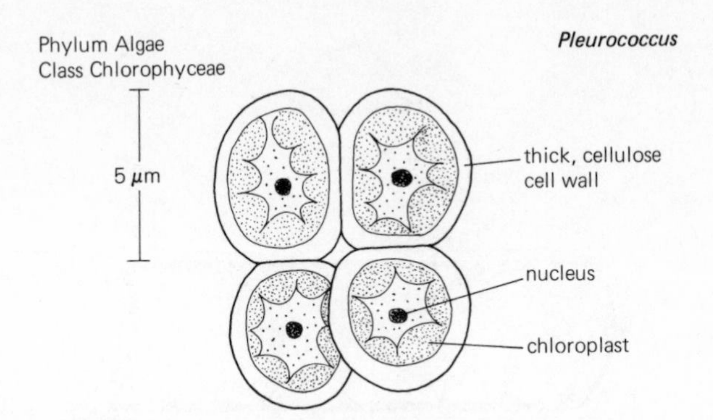

Figure 14 *Pleurococcus* occurs either singly or in small groups. It is a terrestrial form and can be found as a green covering on the north side of trees and walls. Reproduction is by cell division.

---

which group together in clumps. Later, pairs of gametes break away from the clump and fuse. The resulting zygote develops a thick coat and divides to form four 'swarmers'. When released from the zygote coat, the swarmers develop into vegetative cells. The division of the zygote is meiotic, so that, with the exception of the zygote, which is diploid, all other stages in the life cycle are haploid; such a life cycle is described as haplont.

Spirogyra

The cylindrical, vegetative cells of *Spirogyra* are joined to form long, unbranched filaments. Growth of the filaments is intercalary, occuring throughout the filament and not just at the ends of the filament. When the cells reach a certain size, they divide into two by transverse fission. Asexual reproduction is by fragmentation. The filament breaks up into separate cells each of which then grows and divides, giving rise to a new filament. Fragmentation is brought about by the swelling of the cellulose partitions between adjacent cells. In some species changes in turgor pressure are thought to cause the end-walls to bulge, so pushing the cells apart.

Sexual reproduction occurs when the filaments are overcrowded, which in Britain normally occurs in late autumn. Details of the sexual process, which is anisogamous, are shown in Figure 15. As in *Chlamydomonas*, all stages in the life cycle, except the zygote, are haploid. When the zygospore wall ruptures the zygote nucleus divides meiotically, so that the filament that develops is haploid.

Class Bacillariophyceae

Diatoms are found in both fresh and marine waters where many species form a major constituent of phytoplankton (the plant part of the plankton). They are either unicellular or filamentous. A unicellular form, *Pinnularia*, is shown in

Phylum Algae
Class Chlorophyceae

Spirogyra

Vegetative cell

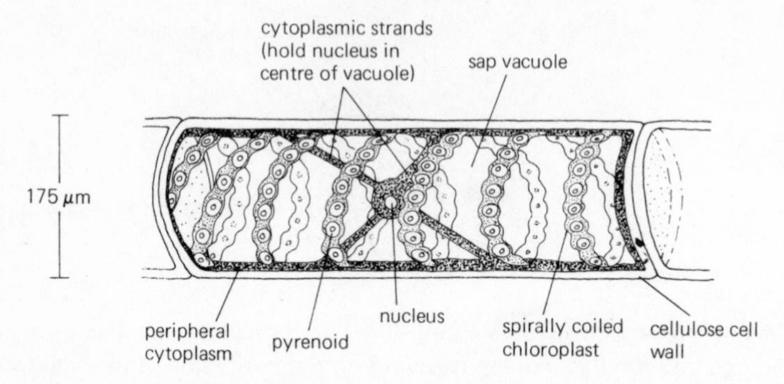

Stages in sexual reproduction

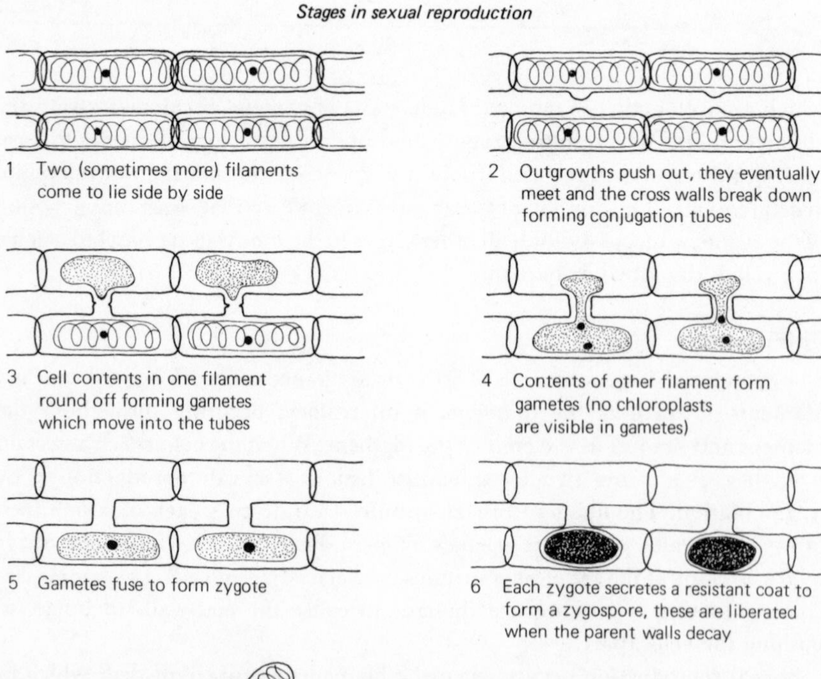

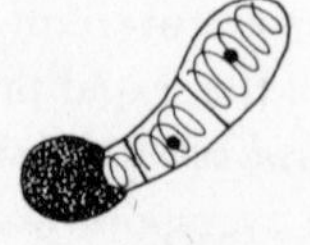

Figure 15 *Spirogyra* has a worldwide distribution. Most species are found floating in stagnant freshwater, although there are some species that live attached to the substrate. The filaments are covered with mucilage which makes them feel very slimy. This is possibly an adaptation to prevent the growth of epiphytes whose presence might interfere with photosynthesis by reducing the amount of light reaching the plant.

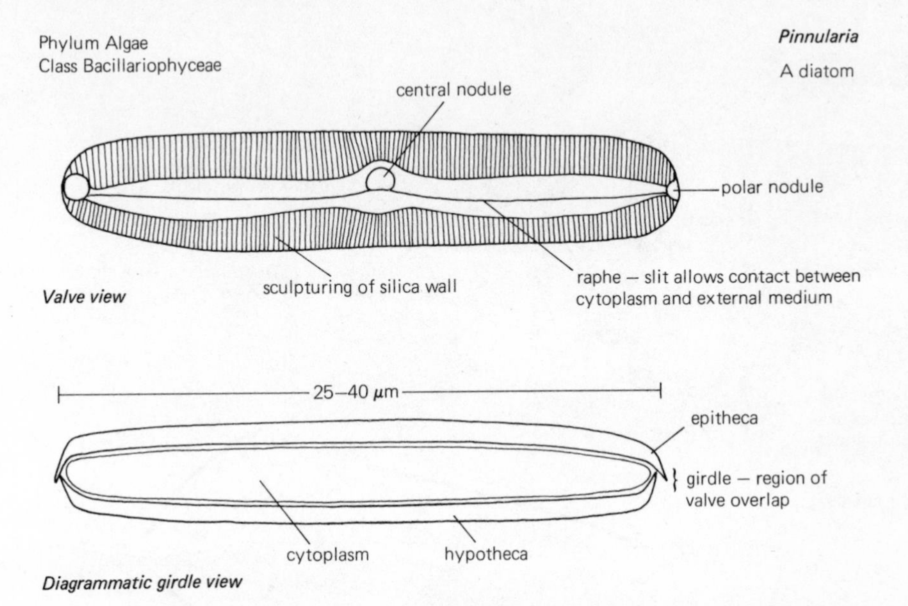

Figure 16 *Pinnularia* is found in marine plankton. They are photosynthetic autotrophs. In cell division the two valves separate, each forms the epitheca of two daughter cells in which new hypothecae form. This leads to a decrease in size in half the progeny; this is generally halted by the cell contents rounding off to form auxospores, these are released from the valves, enlarge and secrete new cell walls thus restoring cells to their normal size.

Figure 16. In addition to chlorophyll, diatoms contain varying amounts of the brown pigment fucoxanthin. Foods stored are oils and a polysaccharide, leucosin.

The most characteristic feature of this group is the cell wall. This is composed of pectin impregnated with silica, and is made up of two halves or valves, one of which, the *hypotheca*, fits into the other, the *epitheca*.

The shells of diatoms form extensive deposits on the sea bed, known as diatomaceous earth. This material is widely used in industrial processes as a filtering agent or as an insulating substance. It is also used as an abrasive in household cleaners and toothpaste.

Class Rhodophyceae

The red colour of these algae is produced by the pigments phycoerythrin and phycocyanin. Most species are marine, although some are found in fast-flowing freshwater streams. Some species are calcareous and certain of these form reef-like structures in tropical waters.

Agar, the gelling agent used in microbiological culture media, is extracted from certain red algae. The species *Porphyra umbilicalis* is gathered and used to

Phylum Algae
Class Phaeophyceae

Fucus vesiculosus

Bladderwrack

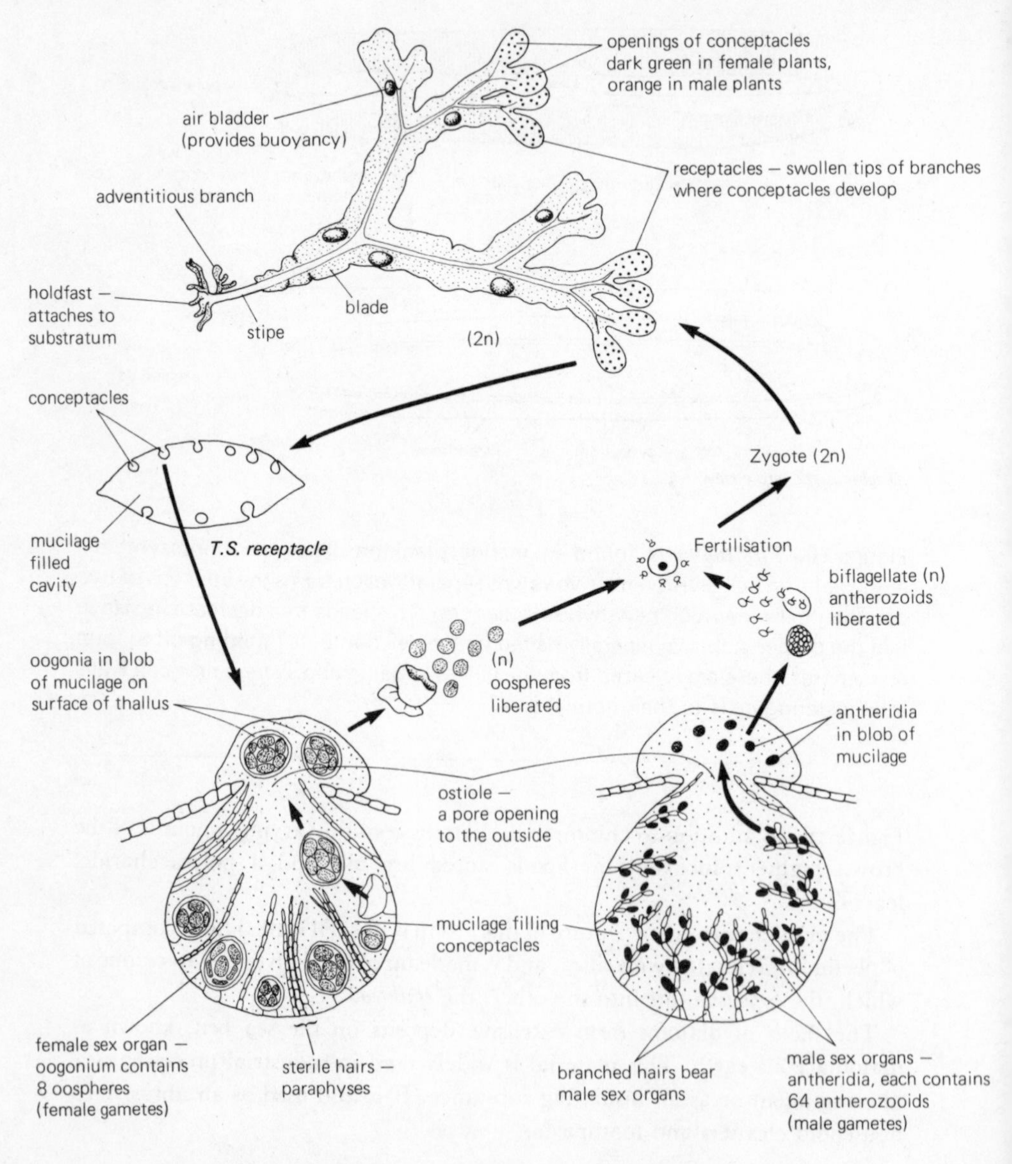

Figure 17 Life cycle of *Fucus vesiculosus*

Fucus vesiculosus is a very common intertidal form. The powerful holdfast and tough but flexible blades enable these plants to withstand buffeting by waves. They also produce large quantities of mucilage, which allows water to be stored, and helps the plant survive the periods of exposure between tides.

Sexual reproduction is oogamous – the female gametes are non-motile and much larger than the more numerous and motile male gametes make laver bread, a favourite breakfast food in Wales. In China and Japan laver is used more extensively and is even cultivated in some areas. Other red algae, notably *Chondrus crispus*, together with many brown algae, are important sources of alginates, the so-called vegetable gelatines, used for thickening soups, setting jellies and emulsifying ice-creams.

Class Phaeophyceae

Nearly all members of this class are marine. They derive their brown colour from the pigment fucoxanthin. This pigment is able to use light from the blue end of the spectrum in photosynthesis. As light of this wavelength penetrates water to a greater depth than light of other wavelengths, this pigment serves to facilitate photosynthesis when the plant is submerged. Food is usually stored as laminarin, a polymer of glucose.

The simplest brown algae are filamentous, but many species exhibit considerable complexity of form, being differentiated into flattened blades with a tough stalk or stipe and an attachment organ, the holdfast, at the base of the stipe. The largest forms are the kelps, some of which reach lengths of 60 metres.

Species of the genus *Fucus* are particularly common on British shores. They are all intertidal but occur at different levels on the shore according to their varying abilities to withstand exposure to air. Stages in the life cycle of *Fucus vesiculosus*, the bladderwrack, are shown in Figure 17.

Fucus vesiculosus

Liberation of gametes in this species is very closely adapted to conditions in the intertidal zone. At low tide the thallus dries out and squeezes the mature sex organs out onto the surface of the plant in a blob of mucilage. As the tide rises, the mucilage and the walls of the sex organs are dissolved, releasing the gametes. The antherozoids swim towards the oospheres, attracted by a chemical secreted by the latter. Eventually one antherozoid penetrates the oosphere and fertilisation occurs. The zygote divides, giving rise to a young thallus, which attaches itself to a suitable substrate by root-like structures called rhizoids. Growth is achieved by the division of a special cell located at the tip of the plant. This special apical cell splits into two at intervals, giving rise to dichotomous branching.

QUESTIONS

1. Describe how the following species are adapted to their environments: a) *Pleurococcus*, b) *Fucus vesiculosus*.
2. Write a brief account of the economic importance of the algae.
3. By reference to one named example in each case, distinguish between isogamy, anisogamy and oogamy.

5 Phylum Fungi

While many fungi, such as mushrooms, toadstools and moulds, are quite widely known, the diversity and importance of this very large group are seldom fully appreciated. There are more than 40000 species of fungi. They live in a very wide variety of habitats throughout the world, the most important being the soil, dead wood and the tissues of living plants. Most species are terrestrial but some, for example the moulds found on dead and wounded fish, are aquatic.

The vegetative plant body or *mycelium* is composed of numerous fine filaments or *hyphae*. These may be *septate*, that is divided into cells by cross walls, or *aseptate*, lacking cross walls; the term *coenocytic* also describes the latter condition. The cells of septate forms are uninucleate, binucleate or multinucleate according to the species and, in some cases, the stage in the life cycle. The protoplasts of septate forms are connected by cytoplasmic strands, which pass through a pore in the centre of each cross wall. The cell walls of fungi are unusual in that they contain chitin, a substance normally associated with animals.

The occurrence of chitin and the unusual form of the plant body are just two instances of the many ways in which fungi differ from most other members of the plant kingdom. These differences have led some taxonomists to place the fungi in a separate kingdom. Perhaps the most fundamental difference between fungi and other plants is their lack of chlorophyll. This means they cannot manufacture their own food and so must rely on sources of ready-made food (they are heterotrophic not autotrophic). Most fungi are either *saprophytes*, feeding on organic remains, or *parasites*, living and feeding on another living organism, the host.

The hyphae of saprophytic fungi grow through the food or substrate, secreting enzymes which diffuse out and digest the food extracellularly, after which the soluble products of digestion diffuse into the hyphae.

Parasitic fungi are divided into two types, namely *facultative parasites* and *obligate parasites*. The former can live on living organisms and organic remains, whereas the latter are more specialised and can only live on a living host. Obligate parasites feed by means of special absorptive organs – the *haustoria*; these are rarely produced by facultative parasites. Haustoria develop as outgrowths from hyphae which penetrate living host cells and absorb nutrients. Obligate parasites do not usually kill their hosts, unlike facultative parasites, which frequently do and then live as saprophytes on the remains.

The diffuse structure of the mycelium is ideally suited to both the parasitic and the saprophytic way of life. The hyphae spread out through the host tissue, or the substrate, making contact with very large areas of food.

Not all fungi are parasites or saprophytes; some live in a permanent symbiotic union with another living organism. Lichens are an example of such an association, involving a fungus and an alga.

Apart from a dependence on ready-made food, fungal nutrition resembles that of animals in other ways; many require an external supply of certain vitamins, and in most species glycogen is the main carbohydrate store – not starch as in most plants.

Reproduction is by means of spores and is generally very prolific. Spores are either sexual or asexual in origin. Except in the most advanced basidiomycetes (see classification), asexual reproduction is the more important means of spreading the species.

In fungi, as in other organisms, sexual reproduction involves the bringing together of two gamete nuclei, which ultimately fuse forming a zygote nucleus. The two nuclei may be derived from the same mycelium or from two different mycelia of compatible mating types (strains). In the first case, where the fungus is self-fertile, it is said to be *homothallic*. In the second case, where the mycelium is self-sterile, the fungus is described as *heterothallic*. The situation is further complicated by the fact that, in some fungi, the fusion of gamete nuclei occurs as soon as they are brought together in the same cell, whereas, in the higher fungi fertilisation is delayed until a mycelium, whose cells each contain two nuclei of different, compatible strains, has been established. Such a mycelium is termed a *dikaryon*, and the way in which this is formed, together with the way in which fertilisation eventually occurs, is illustrated in the life cycle of *Agaricus campestris* (see Figure 23).

CLASSIFICATION

Fungi are divided into three main classes on the basis of the septation of the mycelium and the character and development of the spore-producing organs – the *sporangia*.

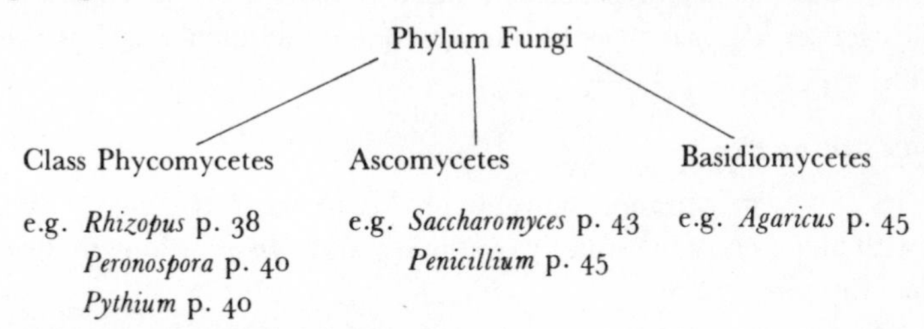

Ascomycetes and Basidiomycetes are often referred to as the higher fungi because of their more complex structure.

ECONOMIC IMPORTANCE

Fungi have a very considerable influence on man's affairs. This is hardly surprising considering the enormous range of food materials they utilise and the very large numbers in which they occur.

Parasitic fungi are responsible for the majority of all plant diseases; they also cause disease in man and other animals. Fungal infections of man are called mycoses; such diseases, although often highly contagious and irritating, rarely cause serious illness.

Saprophytic fungi are responsible for the spoilage of a very wide variety of stored goods such as timber, food, textiles, leather, rubber and even plastics and oil.

On the other hand, many activities of fungi are beneficial. Saprophytic fungi in soil play a vital role in humus formation, especially in acid soils where bacterial activity is low. Many fungi are widely used in manufacturing processes such as the production of industrial alcohol, wines, beers, spirits, antibiotics, cheese-making, baking and many others. In addition the special, delicate flavour of many fungi, notably truffles, morels and field mushrooms, is highly prized in the kitchen.

The following survey of the main classes of fungi includes descriptions of many economically important species.

Class Phycomycetes

This is the smallest class and contains only about 1400 species. They have a relatively simple structure and are generally considered to be the most primitive fungi. Many species are aquatic, and swimming zoospores are encountered in many forms – dependence on water is very often a primitive feature.

The mycelium in Phycomycetes is usually aseptate. The sporangia are relatively simple structures and are not usually grouped together into complex fruit-bodies, but instead are borne singly, or more rarely, in small groups.

Apart from their relatively simple structure, it is generally recognised that the orders within this class have very little in common, and in some classification schemes this class has been replaced by three or more separate classes.

The life cycle of *Rhizopus stolonifer*, the common bread mould, is illustrated in Figure 18.

Rhizopus stolonifer

This species is a very common saprophyte. Members of the genus *Rhizopus*, together with members of the equally common genus *Mucor*, belong to the order Mucorales (the pin moulds).

R. stolonifer derives its name from the stout aerial hyphae, known as stolons, which arch over the substrate before touching down and producing a tuft of root-like, feeding hyphae (rhizoidal hyphae) and a group of sporangiophores (special hyphae that bear the sporangia). This fungus is heterothallic, so plus and minus strains must meet before sex organs (gametangia) will form. The zygospore, formed as a result of sexual reproduction, requires a rest period of several months before it will germinate.

This species and other members of the same genus are quite widely used in the

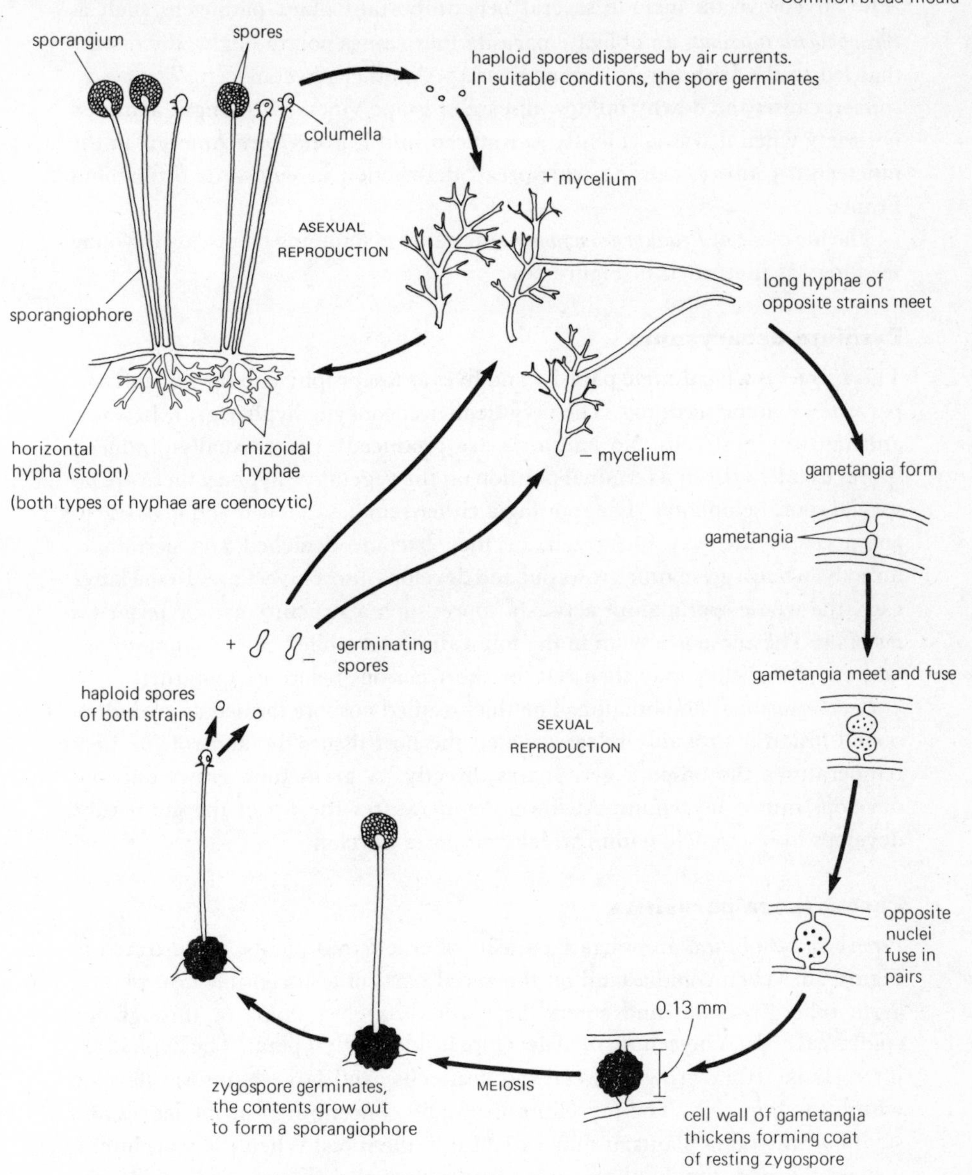

Figure 18 *Rhizopus stolonifer* is a very common saprophyte found living in soil and on damp organic material such as fresh dung and moist bread. This species is also the cause of a serious disease of strawberries in transit, and the soft rot disease of stored sweet potatoes.

Some species of *Rhizopus* can invade the human nervous system often with fatal results.

manufacture of lactic (2-hydroxy propanoic) acid, fumaric (*trans*-butenedioic) acid and the drug cortisone.

*

The Phycomycetes include several very important plant pathogens such as *Phytophthora infestans*, an obligate parasite that causes potato blight, the disease that led to the Irish potato famine in 1845. Another phycomycete, *Plasmopara viticola*, causes the downy mildew disease of grape vines. This fungus achieved notoriety when it was accidently introduced into Europe from America in the nineteenth century, causing widespread destruction in vineyards throughout France.

The life cycle of *Pythium debaryanum*, the cause of damping off disease in young seedlings, is illustrated in Figure 19.

Pythium debaryanum

This species is a facultative parasite and lives as a saprophyte in the soil and as a parasite in young seedlings. The very slender coenocytic hyphae grow between and into the host cells. No haustoria are produced. The asexually-produced spores usually arise in a terminal position on the vegetative hyphae; there are no special sporangiophores. The sporangia either remain attached and give rise to zoospores in the way illustrated, or they become detached and germinate directly, when a germ tube grows out and develops into a mycelium. In the latter case, the whole sporangium acts as a spore; such a structure can be termed a *conidium*. The zoospores swim in the soil water for a while before coming to rest and encysting; they may then rest for short periods before germinating.

P. debaryanum is homothallic. The thick-walled oospore formed as a result of sexual fusion is probably released when the host tissues die and rot. At high temperatures the oospore germinates directly. A germ tube grows out and develops into a mycelium. At lower temperatures the tip of the germ tube develops into a vesicle within which zoospores develop.

Peronospora parasitica

Peronospora parasitica, an obligate parasite of cruciferous plants, is illustrated in Figure 20. When conidia land on the aerial parts of a susceptible host plant a germ tube grows out and enters the plant through a stoma or through the epidermal cells. A mycelium then develops inside the host plant. The hyphae are intercellular (they grow between the host cells) and large haustoria develop within the host cells. The mycelium is usually concentrated in the leaves and stem but, in young plants, it may extend into the roots. When the mycelium is well-established, conidiophores grow out through the stomata, and conidia are produced in very large numbers.

Some strains of *P. parasitica* are homothallic while others are heterothallic. The sex organs are often produced in the deeper parts of the host plant. The thick-walled oospores are released when the host dies and rots at the end of its growing season. The oospore undergoes a rest period before germinating.

*

In the three Phycomycetes just described and illustrated it should be noted

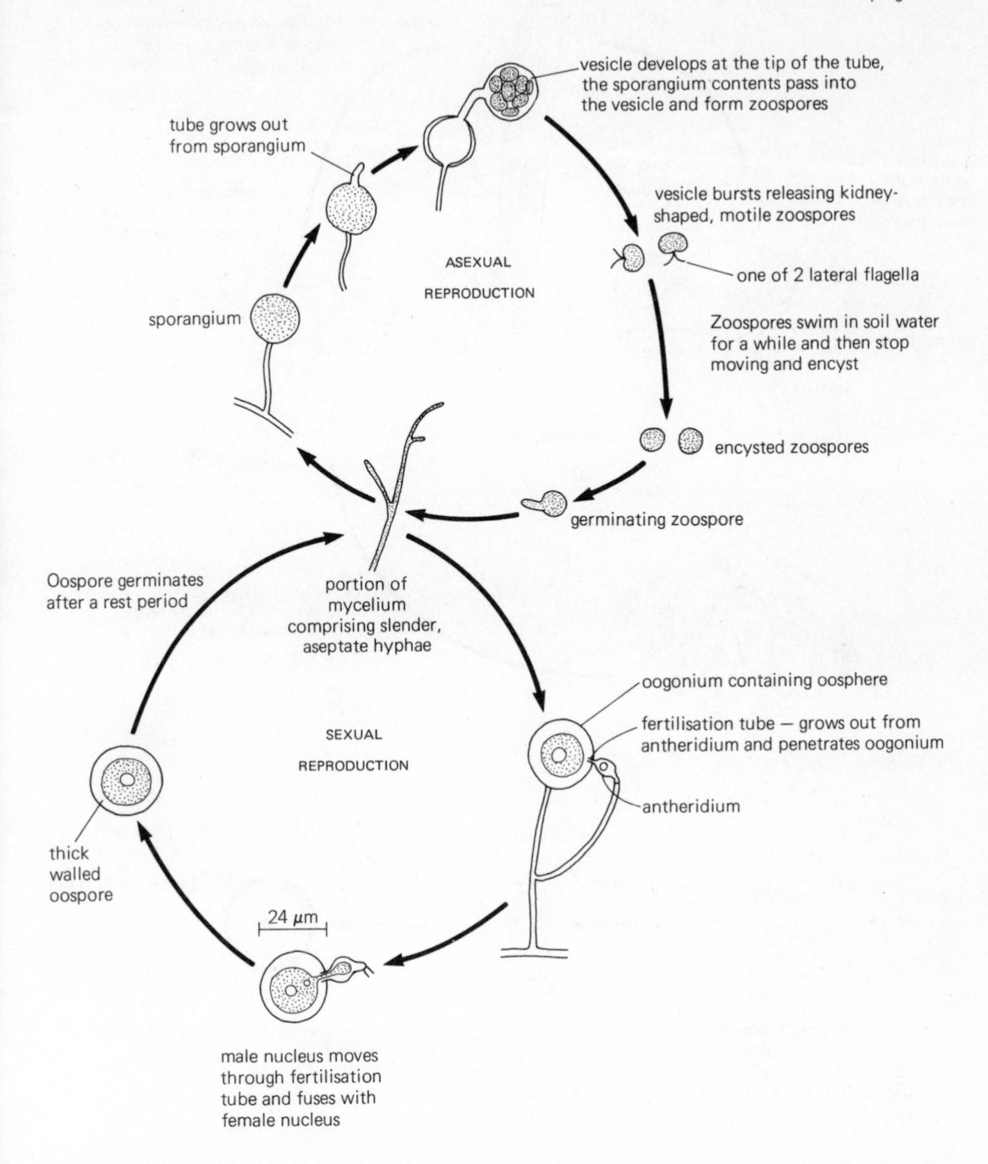

Figure 19 *Pythium debaryanum* is the facultative parasite that causes damping off disease of young seedlings and cuttings grown in overcrowded or excessively wet conditions. It is also common in wet soil where it lives as a saprophyte. The mycelium can penetrate the seedlings and cuttings of a large number of plant species. The fungus lives parasitically on the host, eventually killing it and living saprophytically on the remains.

First signs of infection are watery spots on the stem at soil level which darken, and eventually the seedling may collapse.

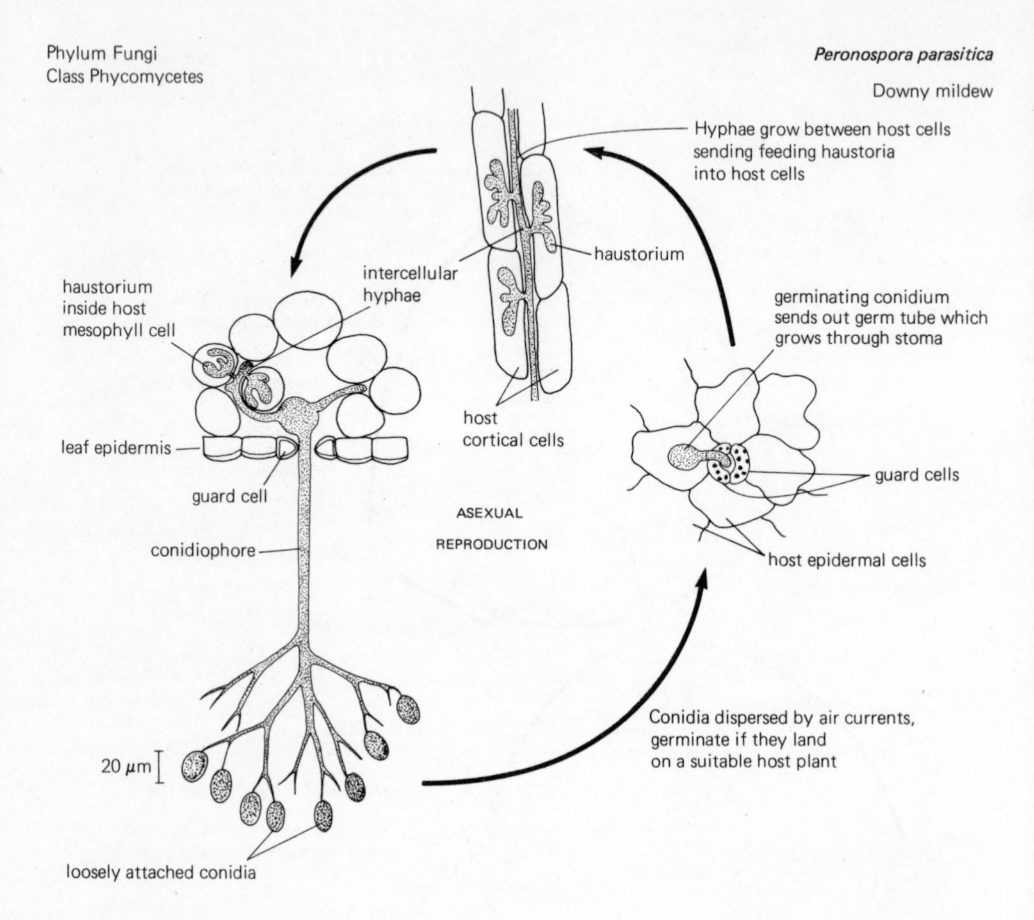

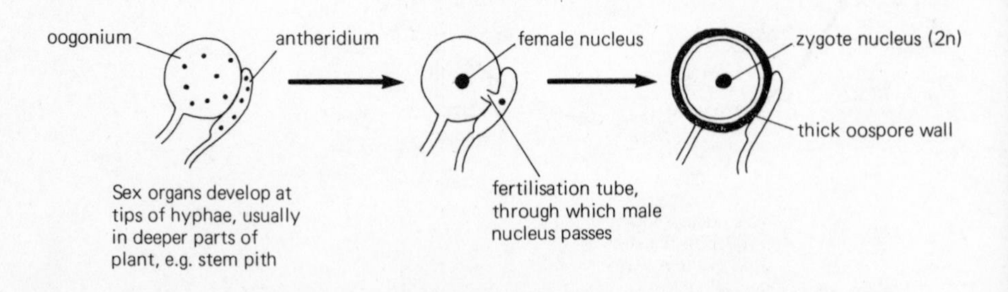

Figure 20 *Peronospora parasitica*, an obligate parasite, causes downy mildew disease of many cruciferous plants, especially Brussels sprouts, cabbages, cauliflowers, swedes, turnips and wallflowers. Diseased plants can generally be recognised by yellow or light green patches on the upper surface of the leaves, and white, furry patches of conidiophores on the lower leaf surface and sometimes on the stem; the latter is often swollen and distended.

Spraying with a copper fungicide is one effective method of controlling this disease; another is to rotate the crops, excluding cruciferous plants for one or more years, to prevent infection by the overwintering oospore.

that the asexual spores are the main means of dispersing the fungus, while the sexual spores are well-protected, resistant spores, that enable the fungus to survive conditions unfavourable for growth.

Class Ascomycetes

This is the largest class containing some 15500 species. They are distinguished from other fungi by their sac-like, spore-producing organ, the *ascus*, within which the sexual *ascospores* develop. Typically, eight ascospores form within each ascus. In most species the asci are grouped together in special fruit-bodies or *sporophores*. The hyphae are usually aseptate and are quite often bound together by vegetative fusions forming a three-dimensional network of fungal tissue known as *pseudoparenchyma*.

This class includes large numbers of plant pathogens. To consider just one genus, *Ceratocystis*, one species, *C. ulmi*, causes Dutch elm disease, while other species cause wilt disease in coffee and rubber, and black rot disease of sugar cane.

Ascomycetes also cause several human diseases, such as thrush, athlete's foot and ringworm. The two last-named diseases are caused by several different species, all of which are able to digest keratin, the tough protein in skin, nails and hair.

Two particularly important and well-known Ascomycetes, *Saccharomyces cerevisiae* (baker's or brewer's yeast) and *Penicillium* are illustrated in Figures 21 and 22.

Saccharomyces cerevisiae

Although often called cultivated yeast, this species does in fact occur in the wild, especially in the sugary liquid on the surface of ripe fruit. Indeed, wild yeast plays an important part in fermenting wine, different strains helping to impart distinctive flavours to different wines.

Unlike other fungi, yeasts are unicellular. This is thought to be an adaptation to life in a sugary, liquid medium, as other fungi, for example *Mucor*, grown in such conditions will assume a unicellular form.

Class Basidiomycetes

This class contains some 15000 species and includes most of the large and conspicuous forms found in fields and woods, such as puff-balls, bracket fungi, stinkhorns and toadstools. It also includes two orders of economically important microfungi, the Ustilaginales or smut fungi, and the Uredinales or rust fungi, both of which cause disease in a wide variety of crops, especially cereals.

The feature that distinguishes Basidiomycetes from other fungi is the *basidium*, a sporangium that produces sexual *basidiospores* on its surface. In most Basidiomycetes the basidia are grouped together in a special layer, the *hymenium*, within the sporophore. In the Gasteromycetes, the sub-class that includes stinkhorns, puff-balls and earth-stars, the hymenia are not exposed and the

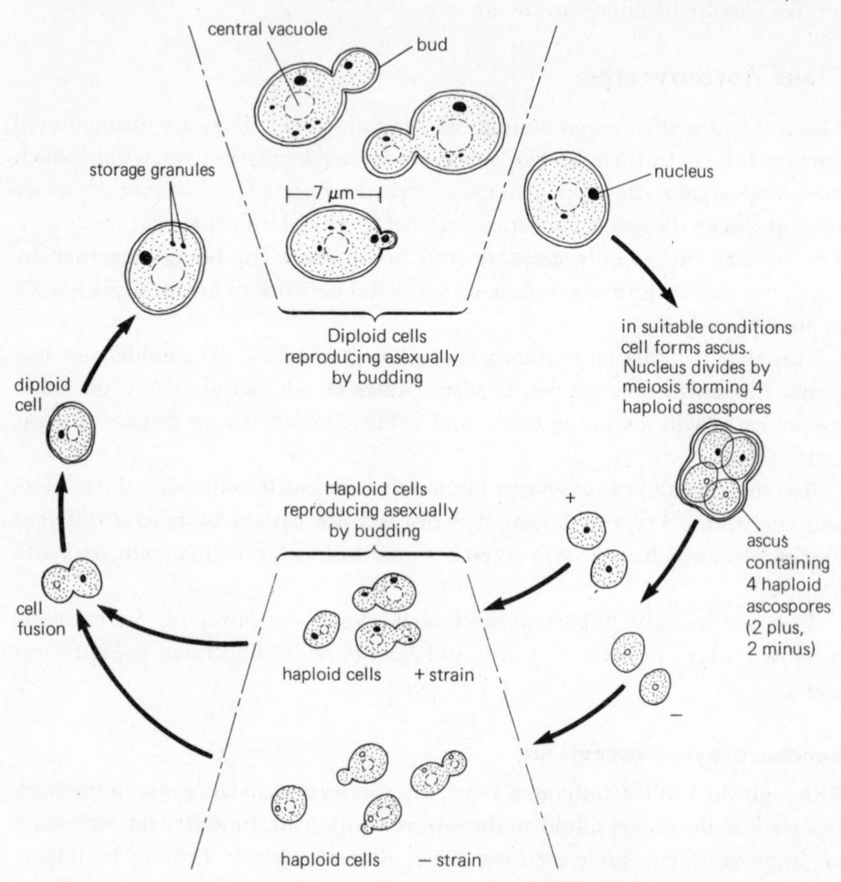

Figure 21 *Saccharomyces cerevisiae*
In a well-aerated sugar solution yeast respires aerobically, growing vigorously and reproducing by budding. In poorly aerated media growth and budding stop and the yeast respires anaerobically, breaking the sugar down to carbon dioxide and ethanol. This process is known as *fermentation* and forms the basis of the brewing and wine making industries.

Yeast is heterothallic and also exhibits alternation of generation, in that both haploid and diploid phases are represented.

---

spores are released by the breakdown of the sporophore. In the Hymenomycetes, the sub-class that includes bracket fungi and toadstools, the hymenia are exposed at maturity, either on gills or via pores, and the spores are released by the basidium which acts as a spore-gun.

The hyphae are normally septate and are often bound together to form pseudoparenchyma.

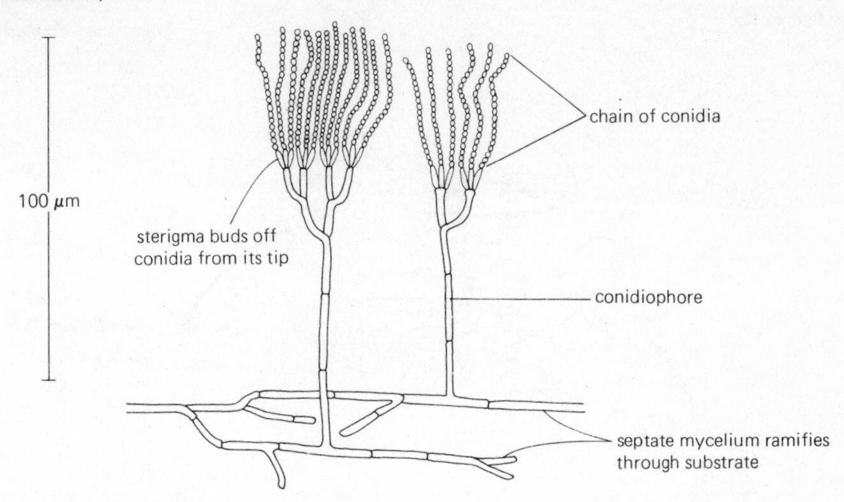

Figure 22 *Penicillium*–blue-green mould
Members of this genus are extremely common saprophytes and occur on a wide variety of substrates, notably citrus fruits. They also cause considerable damage to apples, leather and fabrics in storage, as well as spoiling silage, making it unfit for use as animal feed. On the other hand *Penicillium* is important in many industrial processes such as the production of various organic acids and the ripening of many cheeses. However this fungus is probably best known as the source of penicillin, the first antibiotic to be discovered.

Asexual reproduction by conidia is the only form of reproduction in many species. In the few species in which sexual reproduction does occur it seems to do so only rarely.

It is not generally realised that many of the toadstools found under trees are in fact intimately associated with the tree roots. This association between a fungus and the roots of a higher plant is known as a *mycorrhiza*. The fungal hyphae penetrate the roots and seem to make food from humus available to the higher plant. Many higher plants, including several commercially important timber trees, will not grow as well without their fungal partners. Not all woodland fungi have such desirable effects. The honey fungus, *Armillaria mellea*, a very common woodland species, is a very dangerous pathogen. It invades the cambium of trees through the roots and eventually kills the tree. Another important wood-feeding form is *Merulius lachrymans*, the cause of dry rot in household timber.

The structure and life cycle of *Agaricus campestris*, the field mushroom, are shown in Figure 23. The sporophores of *Agaricus* will only develop from a dikaryotic mycelium (one formed by the fusion of two mycelia of compatible mating types). After the initial fusion of compatible hyphae, the dikaryotic or binucleate condition is spread through the mycelium by the formation of clamp connections through which nuclei pass. Nuclear fusion (fertilisation) occurs in

Phylum Fungi
Class Basidiomycetes

Agaricus campestris
Common field mushroom

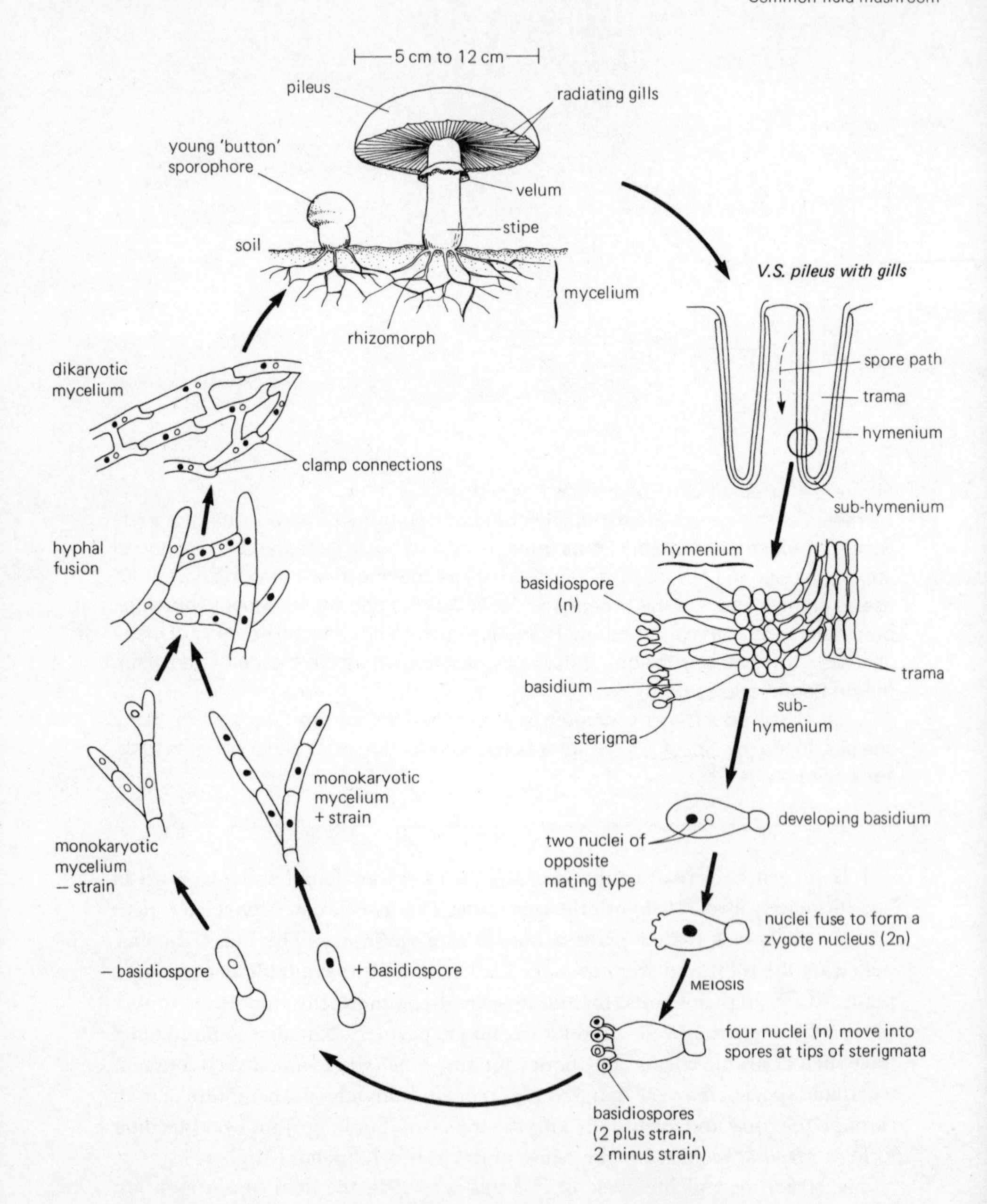

Figure 23 *Agaricus campestris* is a saprophyte, inhabiting soils with a high organic content. The mycelium may live for several years but the fruit-body (sporophore) is a short-lived structure, lasting for only a few days. The sporophores usually appear in early autumn. Food passes to the sporophores through the rhizomorphs, thick threads formed by the union of numerous hyphae.

the basidium, followed immediately by meiosis. The haploid nuclei migrate into spores formed at the tip of the basidium. Each basidiospore is shot off, with an adhering water droplet, to a distance of 0.1 to 1 mm; this is sufficient to carry the spore into the space between the gills. The spore then drops down and, once free of the pileus, is dispersed by air currents. This process is facilitated by the exact alignment of the gills with respect to gravity. When the gills are growing they exhibit positive geotropism. The basidiospores germinate in suitable conditions giving rise to monokaryotic mycelia (each cell contains only one nucleus).

QUESTIONS

1. Explain by reference to named examples the meaning of the following terms: coenocytic, heterothallic and dikaryon.
2. Write a brief account of the economic importance of fungi.
3. With reference to named examples explain how a facultative parasite differs from an obligate parasite.
4. Briefly explain how a named phycomycete and a named basidiomycete differ in the way (i) the species is spread, and (ii) the species survives from year to year.

6 Phylum Coelenterata (Cnidaria)

The Phylum Coelenterata contains a wide variety of simple aquatic animals, many of which are brightly-coloured and delicately beautiful. They include the jellyfish, sea anemones, hydroids, soft and stony corals. Most are marine, but there are a few freshwater forms such as the hydras.

GENERAL FEATURES

Probably the most obvious external feature of the phylum is their radial symmetry about a vertical axis. This means that any cut made vertically through the centre from top to base will result in two identical halves. This kind of symmetry is common in sedentary animals and gives equal access to the environment on all sides (see chapter 1).

The body of a coelenterate is sac-like with a central digestive cavity called the *enteron*, to which there is only one entrance/exit, the mouth. The body wall consists of two layers, an outer ectoderm and inner endoderm. The layers are separated by a non-cellular jelly – the mesogloea, through which cells are able to migrate. The mesogloea becomes particularly extensive in the jellyfish, where it acts in a skeletal capacity for the musculature of the bell. The possession of two body layers is the diploblastic condition.

In coelenterates the bodily arrangement remains essentially simple with no respiratory, circulatory or excretory systems. Instead, exchange with the surrounding water is carried out by diffusion. Although this is a slow process it suffices because the animals are generally small with a large surface area to volume ratio and they also have water circulating through the internal cavities of their bodies.

The simple organisation of the body is reflected in the absence of a definite head – there is no cephalisation, nor is the nervous system centralised, but exists as networks of nerve cells with many cross links. Faster conducting nerve pathways have only been found in the more active species.

A characteristic feature of the phylum is the stinging capsules or *nematocysts*, inside cells called *nematoblasts*. These are found chiefly amongst the ectoderm cells of the tentacles. They are discharged by chemical stimulation or touch, releasing a coiled, sometimes barbed, thread. The thread is ejected at great speed and turns inside out as it goes. Some types end in sharp points which penetrate the prey, allowing a paralysing poison to be injected; others are sticky and grip the prey. Toxins produced by some jellyfish produce rashes, nausea and severe pain in man.

Another feature of the coelenterates is *polymorphism* – having more than one form of individual. The individuals are called *polyps* and there are two basic types, the *hydroid* and *medusa*. The hydroid is a feeding polyp, usually sessile and possessing a cylindrical body, with a mouth at the top, surrounded by tentacles. The medusa is a free-swimming jellyfish with a bell or saucer-shaped body. Depending on the species, one form or the other may be dominant, or absent, or the two forms may alternate in the life cycle.

Many coelenterates show asexual reproduction by *budding* at some stage in the life cycle. Sexual reproduction usually results in a free-swimming ciliated *planula larva* which eventually settles and develops into a new polyp.

CLASSIFICATION

The Phylum Coelenterata is divided into three classes, the Hydrozoa, Scyphozoa and Anthozoa. The classification is based mainly on the degree of importance of the hydroid and medusoid stages in the life history, but also on the arrangement of the enteron and mouth.

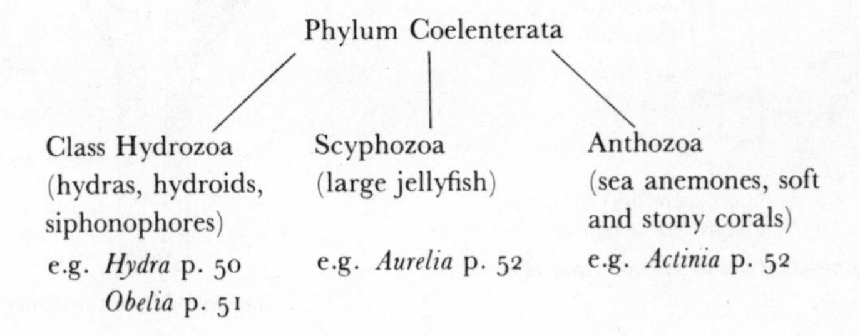

Class Hydrozoa

The Class Hydrozoa contains over 2700 species of hydroids, hydras, medusae and siphonophores. Generally both hydroid and medusoid stages occur in the life cycle, but either may be absent. Most of them only measure a few millimetres in size. The majority are marine but there are a few species of freshwater hydras and medusae.

There are both solitary and colonial examples. Some colonies secrete a chitinous or calcareous skeleton – the *perisarc* around the outside of their bodies. This class shows the most extreme polymorphism in the siphonophores, such as *Physalia physalis*, the Portuguese Man o' War, where various individuals of the colony are modified for feeding, reproduction and even as a gas-secreting float to provide buoyancy.

In this class the enteron remains sac-like in the hydroid form, not divided up into compartments, but in the medusa the enteron is represented by *four radial canals* and *one circular canal*.

Hydra and *Obelia* have been chosen as examples for more detailed study (Figures 24 and 25).

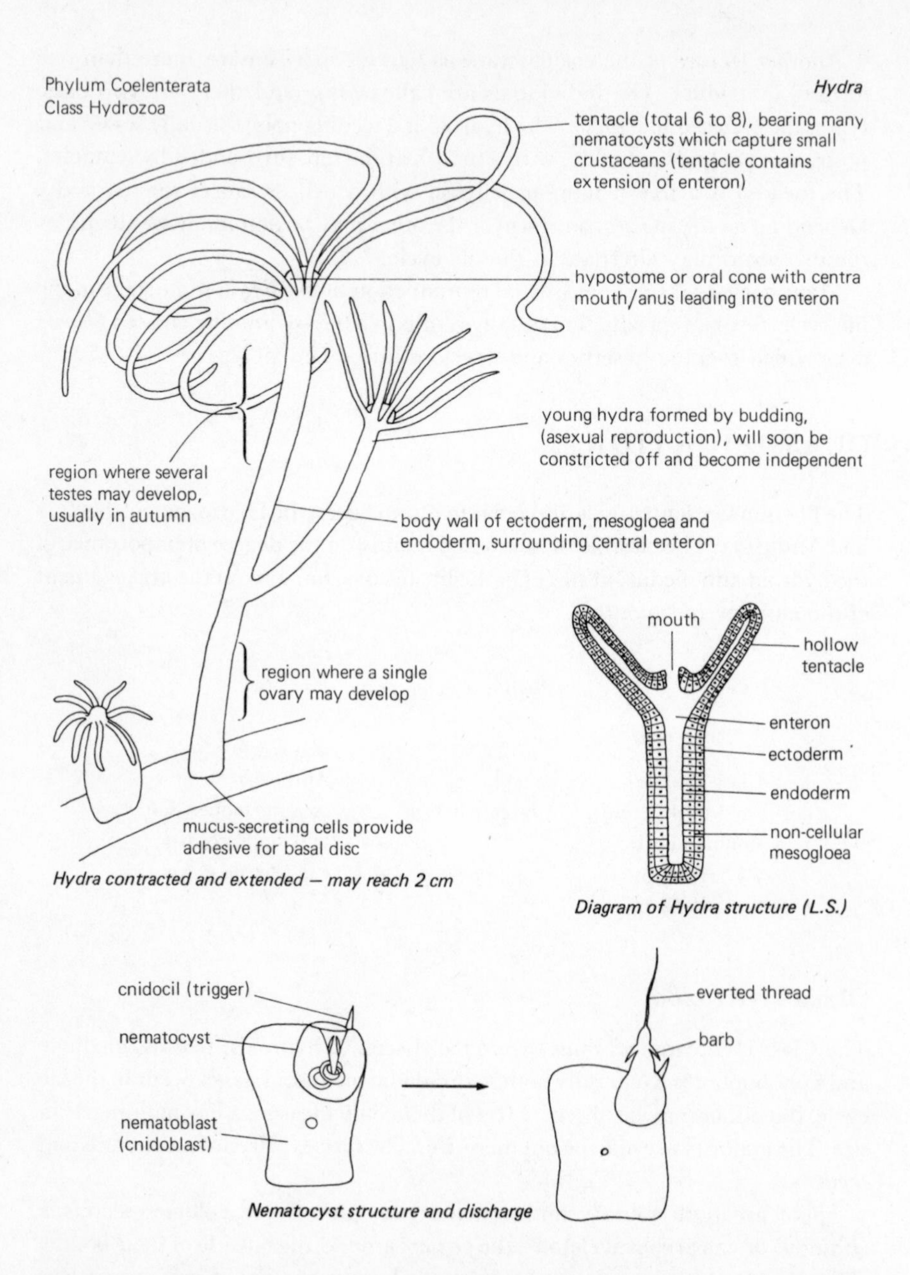

Hydra contracted and extended – may reach 2 cm

Diagram of Hydra structure (L.S.)

Nematocyst structure and discharge

Figure 24 *Hydra* is a genus of solitary freshwater hydroids with no medusoid form. There are many species, differing in colour according to the type of symbiotic unicellular alga present in the endoderm – green *Zoochlorella* or brown *Zooxanthella. Hydra* is found attached to weeds or stones by its sticky basal disc. Asexual reproduction by budding occurs when food is plentiful. The gonads are temporary structures produced only at the time of sexual reproduction.
The embryo is released inside a tough chitinous shell which helps it to survive adverse conditions. *Hydra* rarely moves, but may avoid unfavourable situations by looping, somersaulting or drifting in the water.

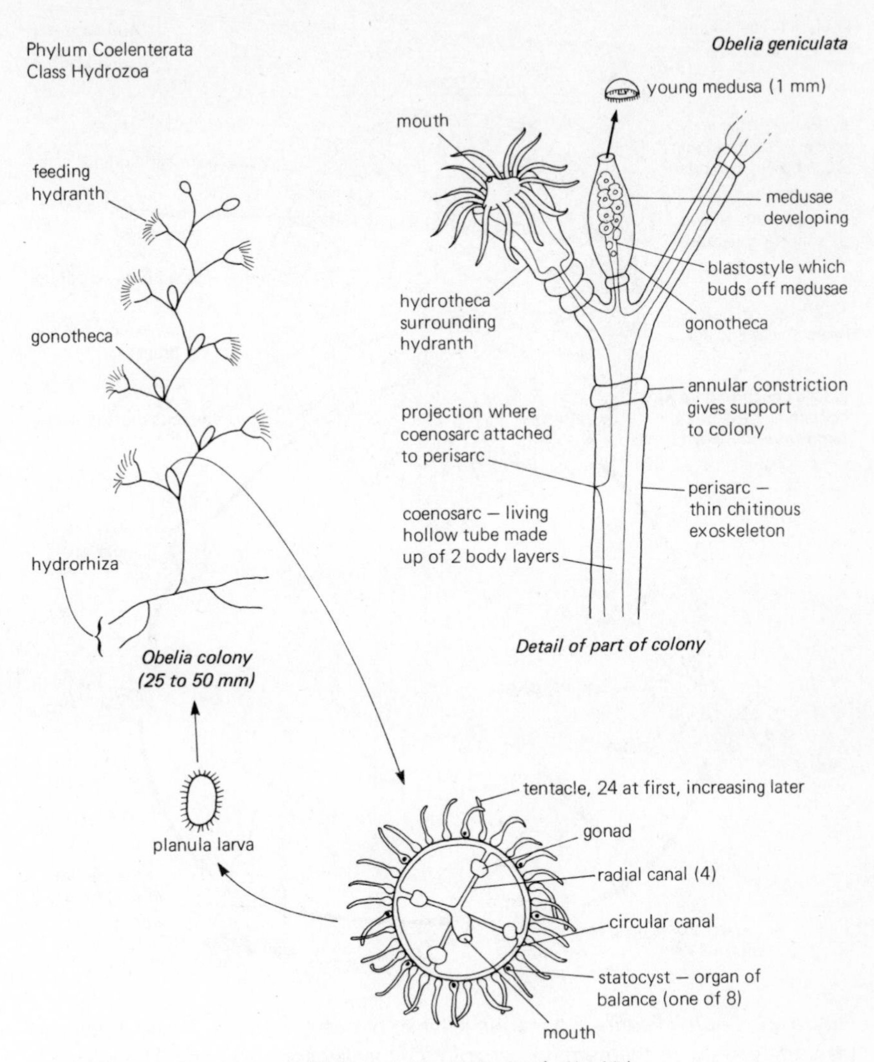

Figure 25 *Obelia geniculata* is a sedentary, marine hydroid colony, found attached to weed and rocks from the inter-tidal zone to a depth of more than 100 m. The colony feeds on small Crustacea caught by the tentacles of the hydranths and since the enteron is continuous throughout the colony, the food caught is shared by all parts. The medusoid stage is budded off from the colony and bears gonads which release gametes into the sea. The fertilised zygote becomes a planula larva which eventually settles and develops into a new hydroid colony. The free-swimming medusa and ciliated larva both help dispersal away from the parent colony.

Although the hydroid and medusoid stages are produced alternately this is not the typical alternation of generations shown by the bryophytes and pteridophytes see chapters 13 and 14, since both phases of *Obelia* are diploid; the gametes being the only haploid part of the life cycle.

Phylum Coelenterata
Class Scyphozoa

Aurelia aurita
Common jellyfish

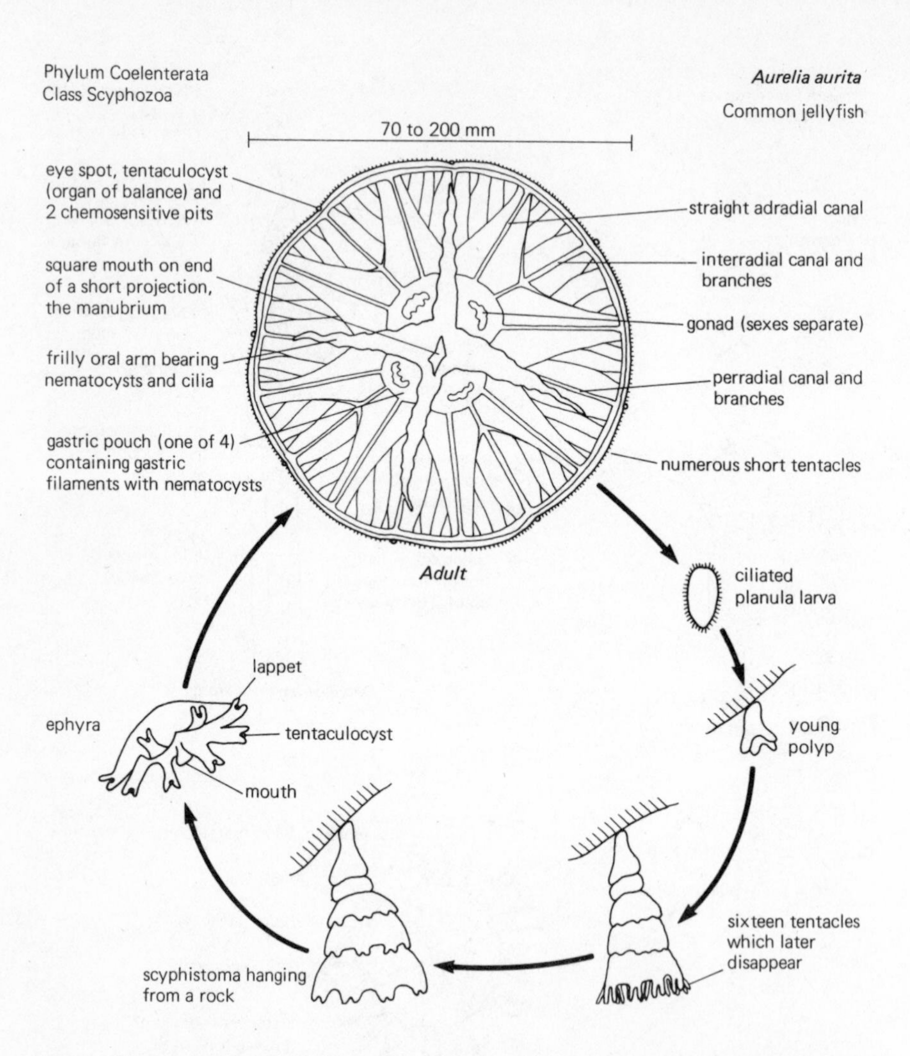

Figure 26 *Aurelia aurita* is a marine jellyfish of coastal waters, where it feeds on plankton caught by the nematocysts of its tentacles and oral arms. The sexes are separate. Sperm are released into the sea and the eggs are fertilised in the oral arms. The zygotes develop into planula larvae which are released and eventually settle. The larva changes into a small hydroid stage – the *hydratuba*. During winter this grows a succession of horizontal buds and is called a *scyphistoma*. Each bud is an *ephyra* – a tiny jellyfish which gradually assumes the adult form.

Figure 27 (opposite) *Actinia* is a common sea anemone of rocky shores, where it is found attached to rocks and weed and feeds on small crustacea. As the tide recedes the anemone contracts using its longitudinal muscles, it is then less liable to mechanical damage or excessive water loss.

The sexes are separate and fertilisation may occur in the sea or inside the body of the female. The zygote becomes a ciliated planula, swimming in the plankton. It eventually settles and develops into a new anemone. Asexual reproduction occurs by longitudinal fission and pieces broken off the basal disc may also become young anemones.

Class Scyphozoa

The Class Scyphozoa contains the large jellyfish, where the medusoid phase is dominant. They vary in size from a few centimetres to over a metre across. The shape of the bell and the number of tentacles varies with the species, but there is a basic four-rayed symmetry with a four-cornered mouth and four *gastric pouches*. From the gastric pouches radial canals lead outwards to a circular canal near the edge of the bell. Other canals then lead back to the bases of the oral arms. Cilia lining the canals beat and circulate oxygenated water and food particles throughout the body. *Aurelia* has been chosen as an example of the group (Figure 26).

Class Anthozoa

The Class Anthozoa includes the sea anemones, soft corals and stony reef-building corals. Here there is only a sedentary polyp, the medusoid form having been suppressed. Members of the group may be either solitary or colonial and with or without a hard covering. Unlike Hydrozoan polyps the enteron is subdivided by septa or *mesenteries*. These are partitions of endoderm supported by mesogloea and ending in groups of *digestive filaments*. On the filaments are nematocysts and cells which secrete enzymes.

Another feature shown by this group is that the mouth opens into a cavity, the *stomodaeum*, lined by ectoderm, which then opens into the enteron. The stomodaeum bears one or two ciliated grooves called *siphonoglyphs*. When the cilia beat, oxygenated water is drawn into the polyp.

Actinia is described as an example of the group (Figure 27).

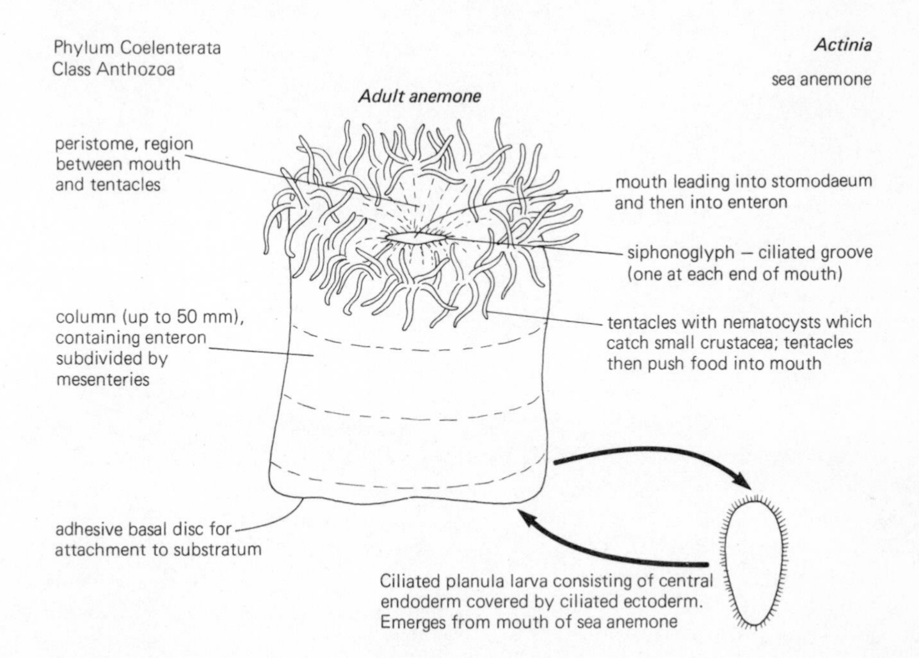

QUESTIONS

1. Explain the meaning of the terms: diploblastic, polymorphism, budding, nematocyst.
2. Using *Hydra* and *Obelia* as examples, explain the meaning of hydroid and medusoid phases and point out the relative importance of each in the life cycle. Why is it misleading to use the term 'alternation of generations' in connection with *Obelia*?

7 Phylum Platyhelminthes

There are more than 5500 species of platyhelminths or flatworms. Nearly three-quarters of these are animal parasites. The phylum is divided into three classes as follows:

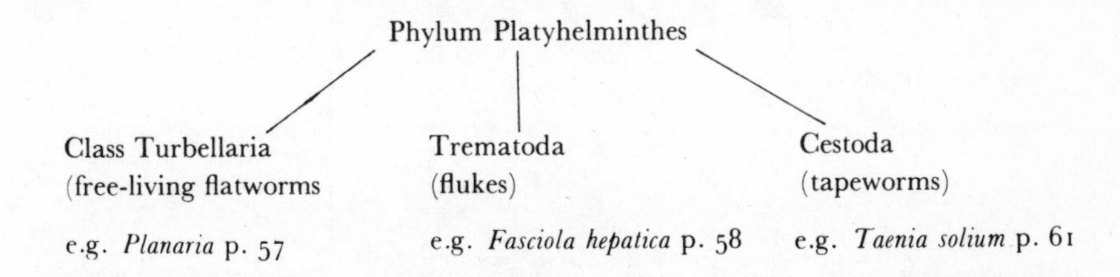

Most turbellarians are free-living. A few species are *commensal*, living in close association with another organism of a different species, on which they depend for food and shelter, without apparently affecting the other organism. Both the Trematoda and the Cestoda consist entirely of parasitic forms.

GENERAL FEATURES

All flatworms are bilaterally symmetrical and dorso-ventrally flattened (from top to bottom). They do not show the serial repetition of parts or metamerism encountered in many other groups. Cestodes are segmented but this is thought to have evolved independently as a means of increasing reproductive efficiency.

In common with all higher animals, flatworms are *triploblastic*, that is, the body is composed of three layers of cells, not two as in coelenterates. The significance of this extra layer is discussed in chapter 1. It is only necessary to say here that the triploblastic condition has made possible a much greater degree of tissue specialisation. This has enabled flatworms to attain an organ system level of organisation – different functions are carried out by a number of different organs working in an integrated way.

Unlike most other triploblasts, flatworms are acoelomate – they lack a body cavity. They also lack a specialised transport system. Presumably their flattened shape provides a sufficiently large surface area to volume ratio to ensure that diffusion alone can supply their transport needs.

The digestive tract is also unusual among triploblasts in that there is only one opening, the mouth, which serves for both ingestion and egestion.

The excretory system consists of numerous excretory cells called *flame cells* because the movement of flagella in the cells resembles the flickering of a candle. The flame cells connect up with a system of branching excretory tubules, some of which open to the outside by excretory pores. This system also serves an osmoregulatory function in many species.

Until fairly recently the outer layer of cestodes and trematodes was thought to be a non-living layer, secreted by the underlying cells. Such a structure corresponds to the outer layer of arthropods and so was called the cuticle. However, recent studies with the electron microscope have shown that the helminth cuticle is in fact a living, metabolically-active layer (Figure 28). In the light of these findings, the cuticle was renamed the *tegument*.

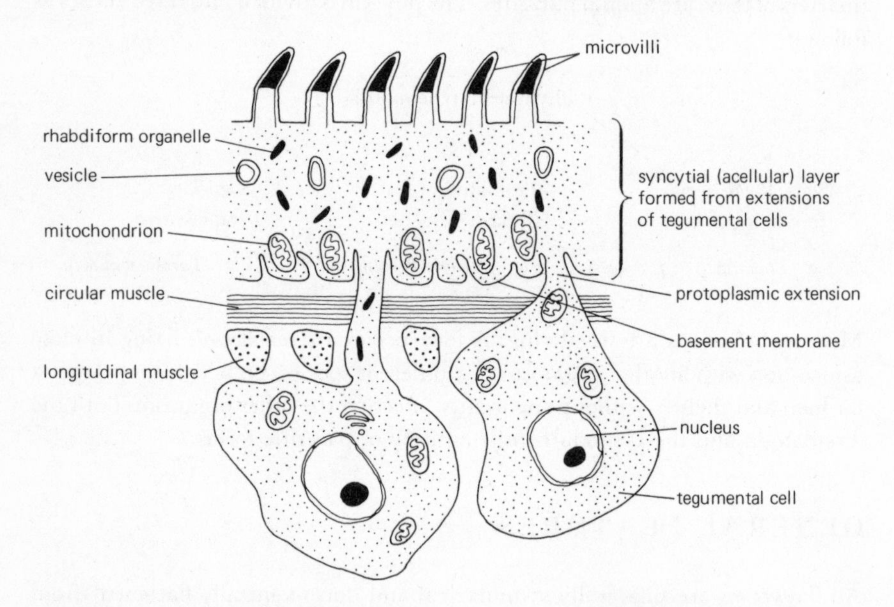

Figure 28 Diagram of a section through the tegument of a cestode as seen with an electron microscope.

Most flatworms are hermaphrodite with well-developed reproductive organs. Parasitic flatworms generally have a rather complex life cycle, passing through one or more larval stages before they reach maturity. In addition, two or more host species are often incorporated into the life cycle. The host in which the parasite becomes sexually mature is known as the *final or primary host*, while the other hosts are called *intermediate hosts*.

Class Turbellaria

With few exceptions turbellarians are free-living. They are mainly aquatic and the great majority are marine. There are a few terrestrial species, some of which attain a very large size (60 cm or more in length), but these are confined to very humid areas.

The body surface is covered with a ciliated epidermis. In some species the cilia are confined to the ventral surface. The beating of the cilia propels the animal along and in so doing creates turbulence – hence the name of the class.

Most turbellarians are carnivores, feeding on small organisms such as protozoa and crustacea. Many will also feed on the bodies of dead animals. There is a well-developed digestive tract in most species. Many flatworms have remarkable powers of regeneration. Quite small pieces can grow into complete animals. The ability of a piece to regenerate depends on its region of origin. Pieces from the anterior region will regenerate much more rapidly, and from much smaller pieces, than those from the posterior region.

The external features of a common freshwater form, *Planaria*, are shown in Figure 29.

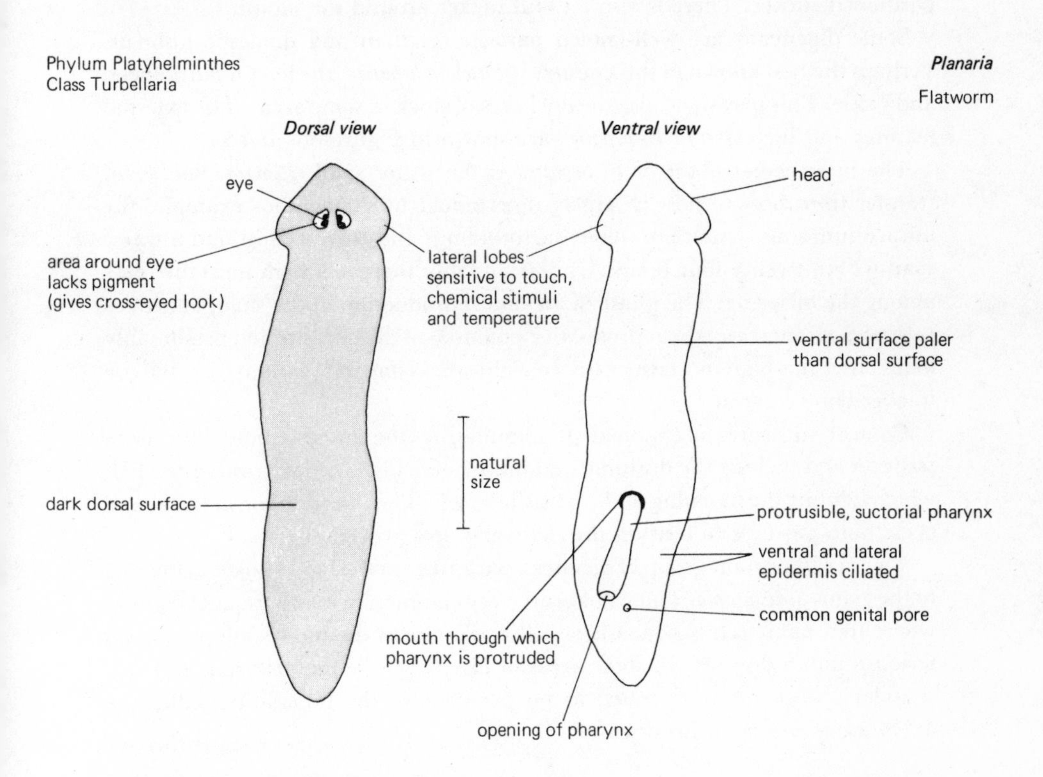

Figure 29 *Planaria* is one of the most common freshwater flatworms and is generally found in slow moving streams and ponds.

The chemoreceptors on the lateral lobes enable planarians to detect the presence of food organisms (mainly protozoans and small crustacea). Gland cells on the ventral surface secrete mucus which helps entangle prey. Once the prey is secured, the pharynx is protruded and enzymes are poured over the food. When partially digested, the food is drawn into the body by the pumping action of the pharynx. Digestion is completed by the amoeboid cells that line the digestive tract.

Although hermaphrodite, cross fertilisation normally occurs. *Planaria* also reproduces asexually by budding or by tearing itself in two, after which each piece regenerates the missing half, a process known as autotomy.

Class Trematoda

These flatworms are entirely parasitic. They have a well-developed digestive tract, and most have prominent suckers. The body of trematodes is covered with a thick tegument, which often contains backwardly pointing spines.

Trematodes fall fairly readily into two main groups as follows:

ORDER MONOGENEA These are mainly external parasites (*ectoparasites*) living on the skin and gills of aquatic vertebrates, notably fish. Their life cycle is usually simple, involving only one host species.

ORDER DIGENEA The adults are usually endoparasites of vertebrates. Their life cycle is normally complex, involving two or more host species. Larval stages are almost invariably found in molluscs. The main adhesive organ is a ventrally-positioned sucker. There is also an oral sucker around the mouth.

Some digeneans are well-known parasites of man and domestic animals. Perhaps the best known in this country is *Fasciola hepatica*, the liver fluke of sheep and cattle. This parasite causes heavy losses of stock in some areas. The external features and life cycle of *F. hepatica* are shown in Figures 30 and 31.

The intermediate host of *F. hepatica* is the water snail, *Limnea*. Successful transfer from host to host is largely determined by chance; for example, the miracidium must hatch into moist surroundings and then it must find a water snail within twenty-four hours. Understandably there is a high mortality rate among the offspring. The phase of asexual reproduction in the snail, known as *polyembryony*, increases the reproductive potential of the parasite and presumably helps offset this high mortality rate. In eight weeks one miracidium can give rise to over 600 cercariae.

Control measures are aimed at eliminating the intermediate host from pastures and include the draining and liming of fields. The latter raises the pH, which inhibits the hatching of the parasite's eggs. The introduction of ducks and geese onto pastures to feed on the snails has also proved effective.

Another important group of digenean parasites are the blood flukes belonging to the genus *Schistosoma*. These flukes are very common in many tropical regions, where they cause schistosomiasis or bilharzia, one of the most widespread and serious human diseases. In some areas of Egypt, where the disease is endemic (regularly occurring), as much as 90 per cent of the population suffer the debilitating effects of this disease.

Class Cestoda

With a few exceptions, adult cestodes are endoparasites in the alimentary canal of vertebrates. Many are important parasites of man. Cestodes lack an alimentary canal. Living as they do, surrounded by pre-digested food, a digestive system of their own is unnecessary; soluble food is simply taken in through the very extensive body surface. As can be seen in Figure 28, the tegument is raised into numerous, hair-like projections, the *microvilli*, which, it is assumed, serve to increase the surface area for absorption.

The body of cestodes is divided into two regions, a head-like *scolex* and a long

Phylum Platyhelminthes
Class Trematoda
Order Digenea

Fasciola hepatica

Liver fluke

Adult (a) digestive tract, (b) external features and excretory system (only half of each of the organ systems is shown)

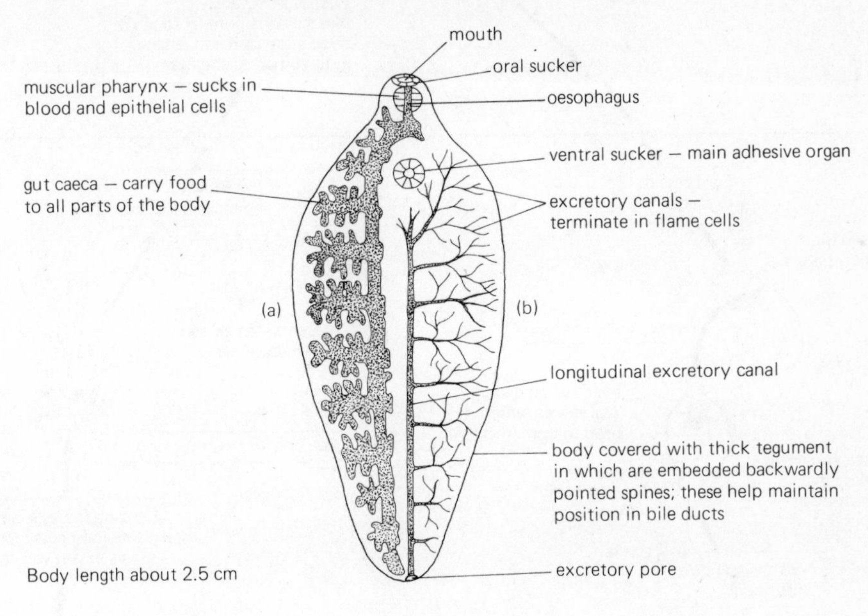

Reproductive system

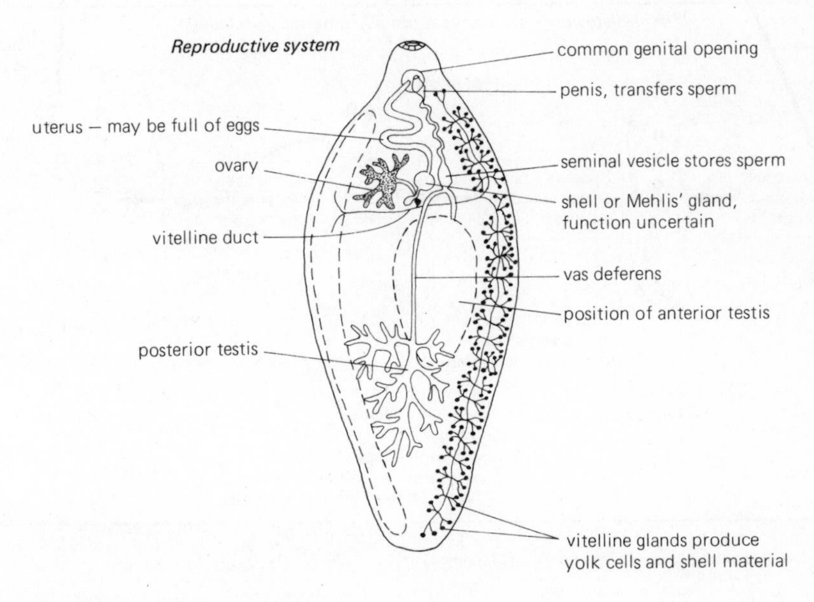

Figure 30 *Fasciola hepatica*
The adult lives in the bile ducts of sheep, goats and cattle. It is occasionally found in man. Such infections are probably acquired by eating the encysted larvae (metacercariae) on uncooked watercress.

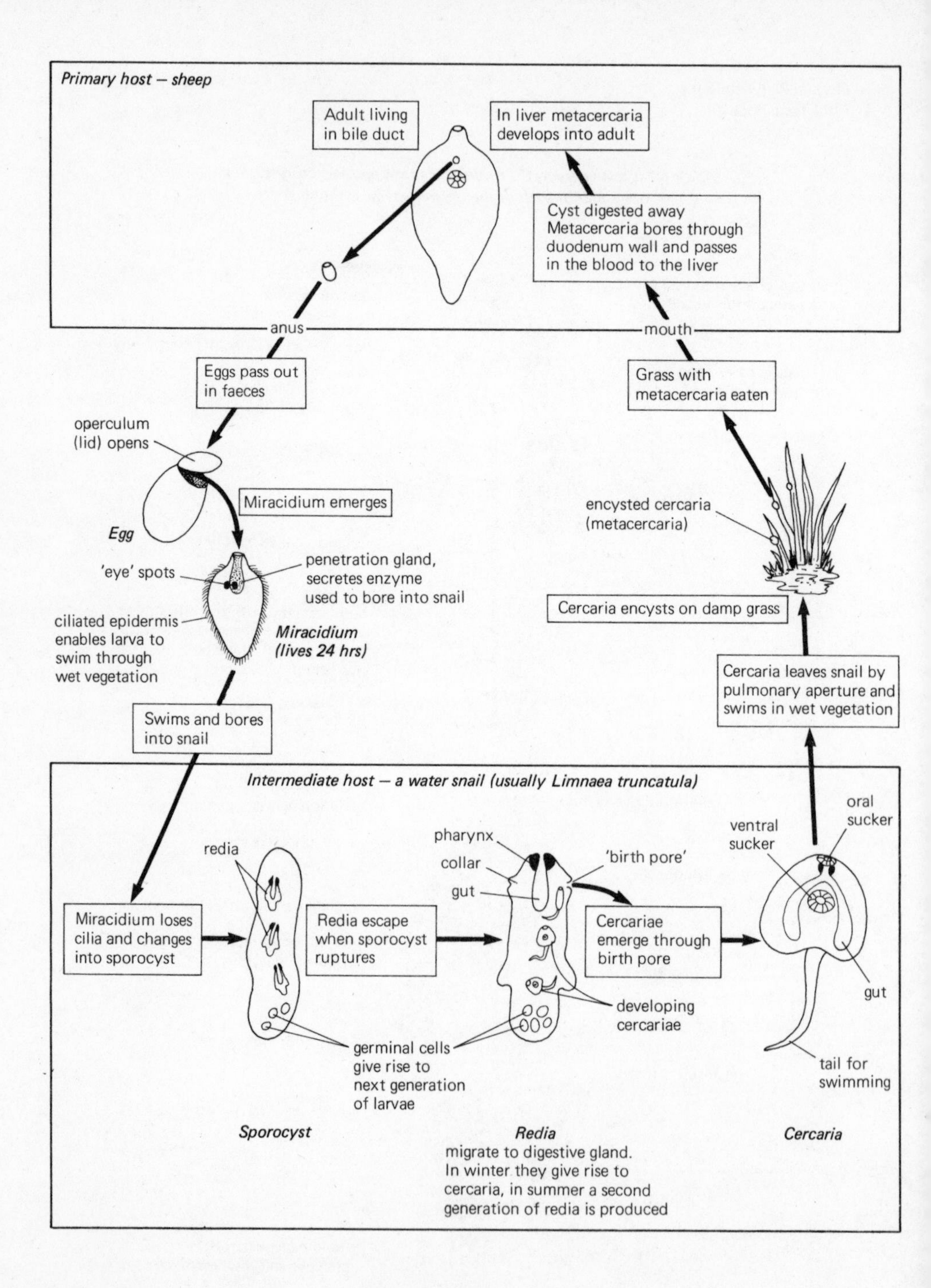

Figure 31 Life cycle of *Fasciola hepatica*
The feeding of the adult flukes and the burrowing of the cercariae can have a very destructive effect on the liver and although sheep can tolerate a few flukes, a heavy infection is usually, very rapidly fatal

tape-like *strobila*. The latter is made up of a number of segments or *proglottids*, each of which contains a set of male and female reproductive organs. The number of proglottids varies from species to species. The dog tapeworm, *Echinococcus granulosus*, is one of the shortest with only three or four proglottids, while the human broad tapeworm, the largest tapeworm found in man, has some 3000 to 4000 proglottids, and has been known to reach lengths in excess of 12 metres.

The scolex bears the attachment organs which are hooks or suckers or both.

The structure and life cycle of *Taenia solium*, the pork tapeworm of man, are illustrated in Figures 32 and 33. The adult lives in the small intestine and feeds by absorbing pre-digested food over its body surface. Although hermaphrodite, the male sex organs mature before the female organs – a condition known as *protandry* – this precludes self-fertilisation. If only one worm is present, as is often the case with *T. solium*, fertilisation occurs when the tape folds back on itself. Fertilised eggs pass into the uterus, which eventually comes to occupy most of the space within the proglottid as the other organs degenerate. Development of the eggs starts in the uterus and by the time the proglottid is shed, the uterus may contain as many as 40000 onchospheres. An *onchosphere* consists of a six-hooked (hexacanth) embryo enclosed within a striated membrane, the embryophore, and a thick shell. Small groups of proglottids are shed at intervals from the end of the tape and pass out with the faeces of the host.

Taenia seems to be a well-adapted parasite in that it has little or no adverse effect on healthy adult hosts; although in children and less healthy individuals the presence of this tapeworm can result in abdominal pains, loss of weight and nervous disorders.

ADAPTATIONS TO PARASITISM

The parasitic helminths exhibit many of the adaptations commonly associated with a parasitic way of life, these include:

1 attachment organs – these enable the parasite to maintain its position in the right part of the host, e.g. the tegumental spines and suckers of *Fasciola*, and the hooks and suckers of *Taenia*;

2 reduction or loss of organs no longer essential for survival. For example, in both *Fasciola* and *Taenia* there is a reduction of sense organs as compared with free-living flatworms. Presumably, living as they do in a relatively uniform environment, there is little need for a complex monitoring system;

3 dispersal mechanisms. Overcrowding of any habitat is undesirable as it reduces the amount of food and space available to the organisms living there. However, in the case of parasites the effects of overcrowding can be particularly serious – as well as increasing competition for nutrients, living space, etc., overcrowding can result in the total destruction of the environment, i.e. the death of the host. There is therefore a particular need for effective methods of dispersal, and so incorporated into the life cycle of many parasites are free-living larval stages which serve to disperse the species. However, this creates a new

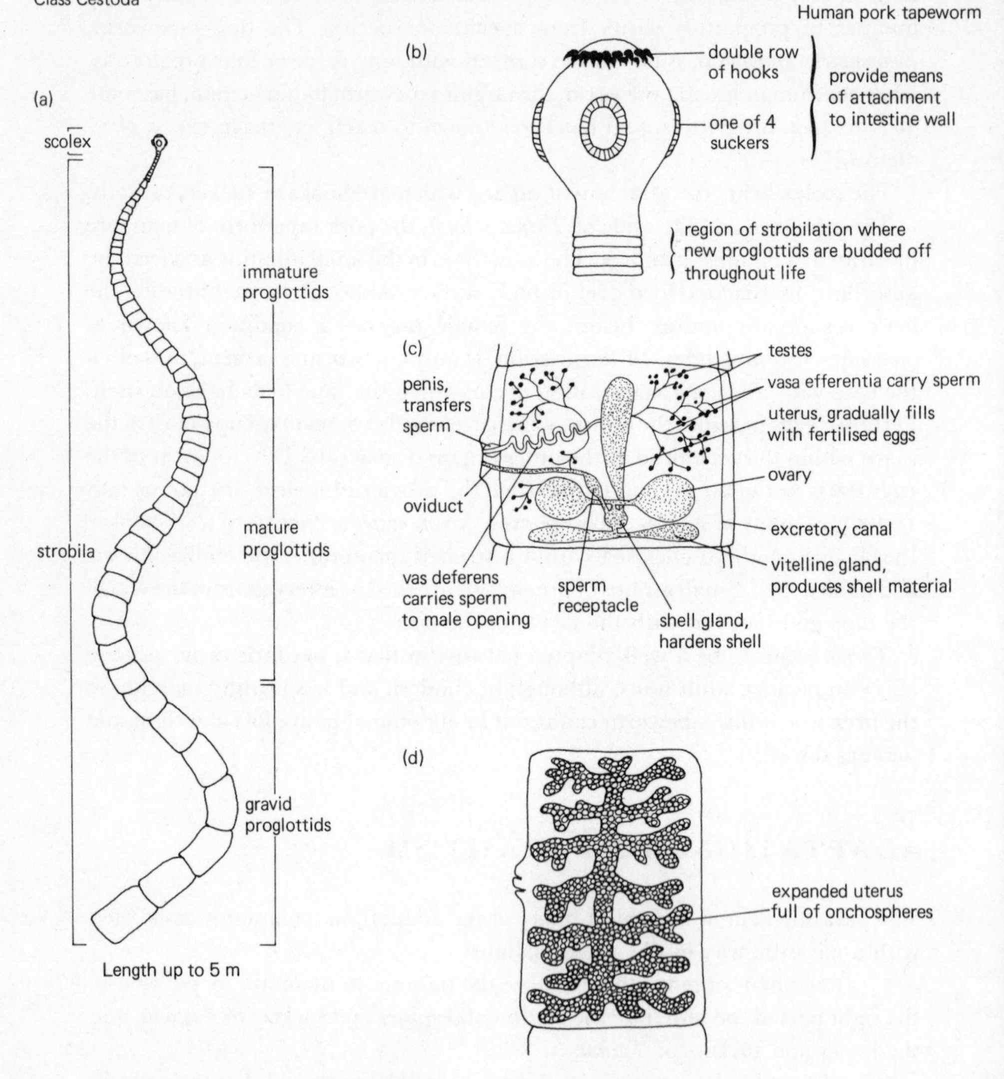

Figure 32 Adult *Taenia solium* (a) whole worm, (b) scolex, (c) mature proglottid, (d) gravid proglottid

difficulty in that the free-living larvae must find another host. In many parasites this difficulty has been overcome by incorporating an intermediate host organism, which facilitates transfer from one final host to another, into the life cycle. It is not difficult to imagine how the close association between man and the domestic pig led to the latter being incorporated into the life cycle of *T. solium*;

4 prolific reproduction. The life cycle of parasites is generally fairly hazardous. The free-living stages are especially vulnerable as there is usually a

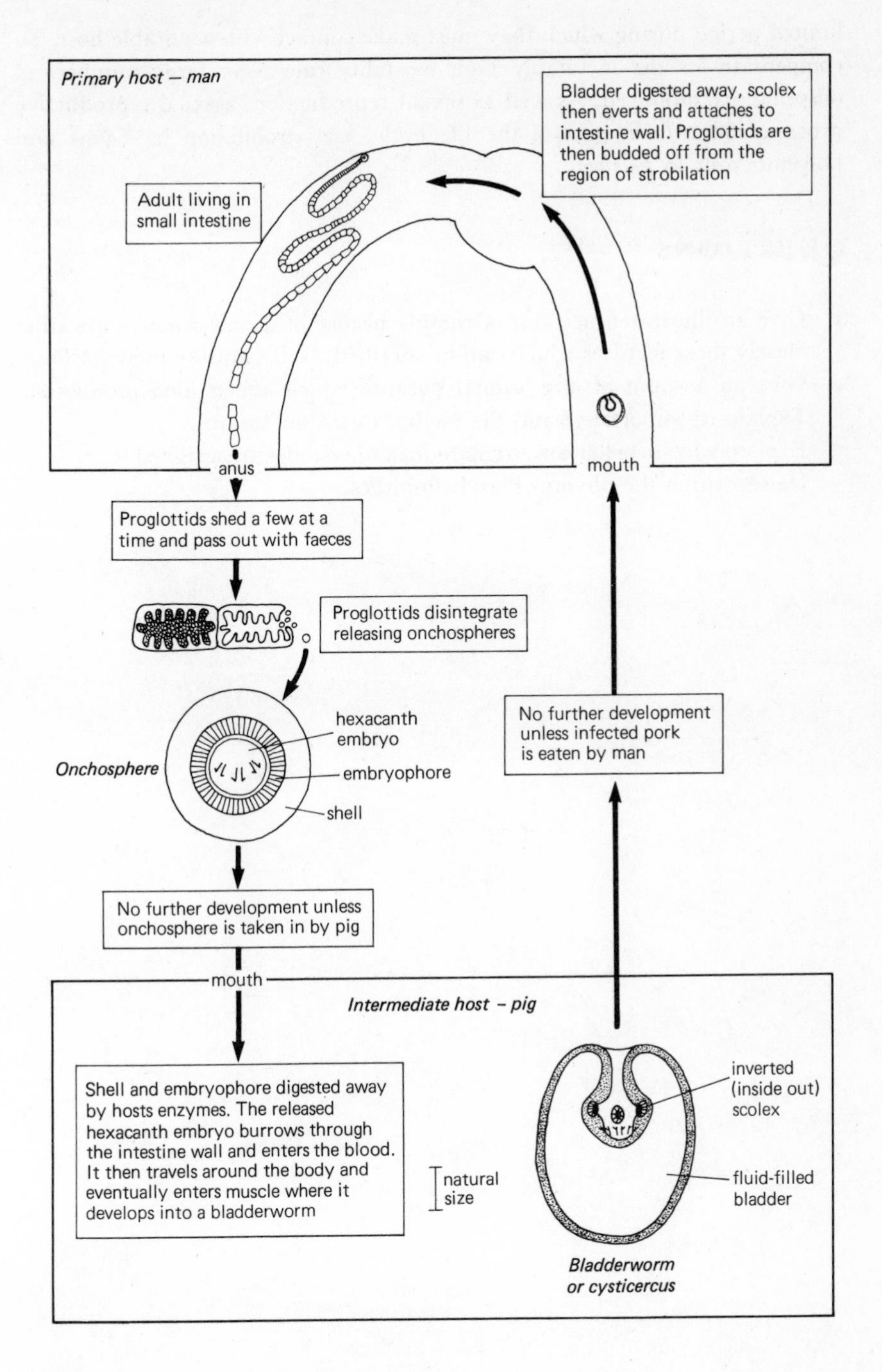

Figure 33 Life cycle of *Taenia solium*
In the past *T. solium* was one of the most common parasites of man but careful disposal of human sewage, improved hygiene in pig farming and inspection of all pork in abattoirs for bladderworms ('measly' pork) have led to the virtual eradication of this parasite in Britain and the United States of America

limited period during which they must make contact with a suitable host. To compensate for the inevitably high mortality rate, very large numbers of offspring are produced. As well as sexual reproduction, asexual reproductive processes often form part of the life cycle, e.g. strobilation in *Taenia* and polyembryony in *Fasciola*.

QUESTIONS

1 Give an illustrated account of the life history of *Taenia solium*, indicating clearly those features which can be ascribed to its parasitic mode of life.
2 Give an account of any animal parasite which affects food production. Explain its importance and the methods used for control.
3 Explain why turbellarians, trematodes and cestodes are assigned to separate classes within the phylum Platyhelminthes.

8 Phylum Nematoda

Apart from the species which are important parasites of man, the nematodes or roundworms are a little-known group. This is probably because the majority are small, inconspicuous forms living in places where they are not normally seen. In fact, with more than 10000 species living in very diverse habitats, they are one of the largest and most widely distributed invertebrate groups. They are found in the sea and in freshwater, in the soil, where as many as 10^{10} per acre have been recorded, and as parasites of plants and animals. They have even been found in such seemingly inhospitable situations as vinegar and hot water springs.

Probably most animal species, including man, harbour nematode parasites sometime during their lives. The majority cause little or no harm to their hosts but a few are responsible for some widespread and serious human diseases; these include hookworms and filarial worms. Hookworms live attached to the intestinal lining and feed by sucking blood, causing anaemia and lethargy and sometimes, in children, mental and physical retardation. One of the best known filarial worms is *Wucheria bancroftii*. This species lives in the lymphatic system and very heavy infections can cause the massive swelling and growth of tissues known as elephantiasis. Both hookworms and filarial worms are restricted to tropical and sub-tropical regions, unlike the pinworm, *Enterobius vermicularis*, a very common but relatively harmless, intestinal parasite of children living in temperate regions.

Nematode parasites are also one of the most important causes of crop damage. They attack a wide range of crops including coffee, cotton, potatoes, sugar beet and tea. It has been estimated that nematodes are responsible for the loss of 10 per cent of all crops planted in the USA.

GENERAL FEATURES

Considering the diversity of their habitats and diet, the external structure of nematodes is remarkably uniform. The smallest nematodes are microscopic, the largest are about one metre long. The principle features of the group are illustrated in the diagram of *Ascaris* (Figure 34). The body is unsegmented and cylindrical, tapering to a point at each end. There are two openings in the digestive tract, a feature roundworms share with all higher animals. The body is covered with a thick, elastic cuticle. It has recently been discovered that the nematode cuticle, like that of helminths, is a metabolically-active, syncytial layer.

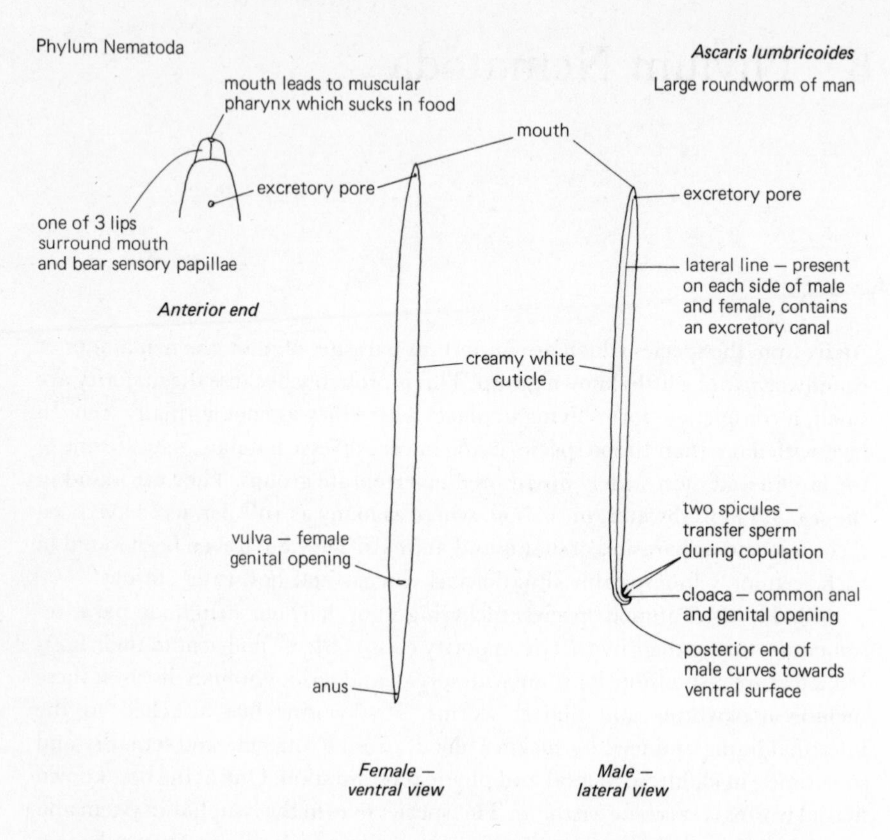

Figure 34 *Ascaris lumbricoides* lives in the small intestine of man and pigs, feeding on intestinal contents. Heavy infections seriously reduce the amount of food available and may completely obstruct the gut.

The females lay thousands of eggs per day. These pass out with the faeces and once outside the body of the host can remain infective for up to six years. A new host is infected by ingesting eggs. Common sources of infection are dirty hands and vegetables contaminated by contact with sewage.

In common with many other nematode parasites, the greatest damage is inflicted on the host during larval migrations. After hatching the larvae burrow through the intestine wall into the bloodstream. They pass to the lungs which they penetrate and eventually crawl up the trachea to the throat where they are swallowed and thus return to the intestine where they complete their development. Larval penetration of the intestine and lung can result in severe internal bleeding.

The body cavity of nematodes develops in a different way from that of most other animals and is accordingly called a *pseudocoelom*. The latter is also unusual in that the hydrostatic pressure of the coelomic fluid is very much higher than it is in other animals.

Beneath the cuticle there is an epidermis and under this a layer of longitudinal muscle arranged in four blocks. These muscles must act against an antagonistic force and, since there are no circular muscles, this is provided by the hydrostatic pressure of the pseudocoelomic fluid. The thick cuticle prevents any increase in the diameter of the body; thus nematodes can only move by alternate bending and contraction of opposite sides of the body, resulting in the peculiar, whip-like movement so characteristic of this group.

The sexes are usually separate and the females are usually larger than the males. The life cycle normally includes six stages – the egg, four larval stages and the adult. The end of each larval stage is marked by the moulting of the cuticle.

QUESTION

1 State what you know of the nematodes and write an account of one named nematode that is known to be harmful.

9 Phylum Annelida

The annelids are segmented worms. The majority are marine animals of shallow water. Some are free-swimming, but others build temporary or permanent tubes or burrows in which they live a sedentary existence. Better known members of the phylum are the earthworms, which also have related species in freshwater. The leeches form a distinct group found as predators or parasites in freshwater or moist soil.

GENERAL FEATURES

The most obvious characteristic of the phylum is the cylindrical shape of the body, which is composed of a series of similar segments, separated from each other internally by septa. At the head end is a *prostomium*, which may bear sense organs and at the tail end is the *pygidium* – the last segment. The rest of the segments are very similar and may bear small bristles called *chaetae* and sometimes paddle-like appendages called *parapodia*.

Annelida are bilaterally symmetrical, triploblastic and coelomate, see chapter 1. The body wall consists of an outer soft cuticle, an epidermis, a layer of circular muscle and then a layer of longitudinal muscle. Inside the muscle layers is an extensive coelom filled with coelomic fluid. Since this fluid is incompressible the circular and longitudinal muscles are able to push against it and act antagonistically (in opposition to each other). The coelomic fluid therefore serves as a hydrostatic skeleton.

In the centre of the animal is a tubular gut also surrounded by muscle layers. The blood is enclosed in blood vessels and is therefore a closed circulatory system. There is a group of valved tubes – the *pseudohearts*, responsible for pumping blood around the body. The principal blood vessels are longitudinal ones but there are also segmentally-arranged lateral vessels.

Some Annelids possess gills, others use their moist skin for gas exchange. The excretory system is made up of segmentally-arranged pairs of organs called *nephridia*.

There is a very simple brain composed of two dorsal *cerebral ganglia*, attached to a *double ventral nerve cord* which passes back down the body to the tail end, expanding into ganglia in each segment.

CLASSIFICATION

The annelids are divided into three main classes. The Polychaeta, or bristle worms, have many chaetae, usually a distinct head and parapodia. The Oligochaeta, which include the earthworms, have few chaetae, a less obvious head and no parapodia. The Hirudinea, or leeches have suckers but no chaetae.

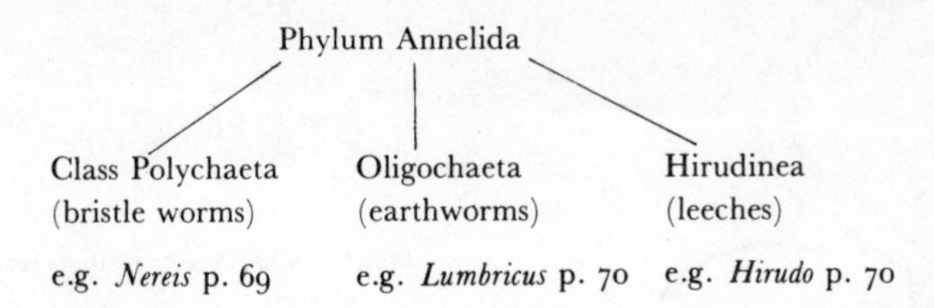

Class Polychaeta

The bristle worms are the largest and least specialised class of the Annelida. They are common seashore animals found under stones, or in tubes or burrows in sand and mud. Earlier classifications divided them into two sub-classes, the Errantia and Sedentaria. The errant polychaetes are the more active ones, usually with more marked cephalisation and greater development of the sense organs, as in *Nereis*, the ragworm. The sedentary types tend to be specialised for life in a burrow, having reduced parapodia and less distinct prostomium. *Arenicola*, the lugworm is a good example. This classification is not a natural one however and does not take into account the evolutionary relationships of the worms.

Nereis will be described as an example of the Polychaeta.

Nereis

The sexes are separate in *Nereis* and at spawning time most nereids metamorphose into active *heteronereids* wth better developed eyes and parapodia. They swim up to the surface layers of the sea, where the gametes are released by rupture of the body wall. The adult worms then die. The zygote develops into a free-swimming *trochophore* larva which gradually elongates and begins to develop segments. The trochophore (or trochosphere) is particularly interesting because it occurs in other groups, including the Mollusca see p. 100, thus suggesting a closer evolutionary relationship between these groups than is indicated by the adult organisms. (See also Figure 35.)

Class Oligochaeta

Earthworms make up the bulk of the Oligochaeta, but the group also includes some freshwater forms, such as *Tubifex*, the blood worm, sold as fish food. The reduced number of chaetae, absence of parapodia and small prostomium are all adaptations one would expect in burrowing animals. *Lumbricus*, the common earthworm is described as a representative of the group (Figure 36).

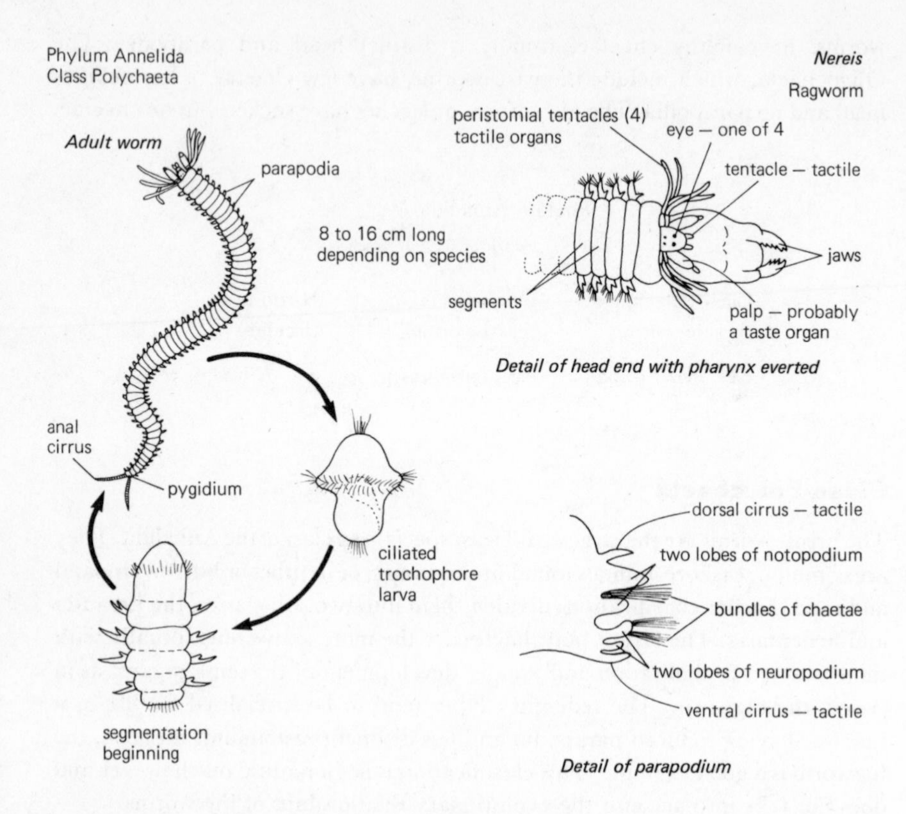

Figure 35 *Nereis*, the ragworm inhabits the intertidal zone of shores and estuaries. At low tide it burrows into sand or mud, but can also swim and crawl. There are several species of nereids, many are active carnivores with well-developed jaws, but some feed on seaweed, or detritus which they trap in mucus inside the burrow. When the pharynx is everted through the mouth, the jaws can seize food, which is then swallowed when the pharynx is inverted again. Inside its burrow gentle undulations of its body bring the worm a current of oxygenated water.

Figure 36 (opposite, above) *Lumbricus terrestris* is an earthworm of moist, non-acid soils. It feeds on organic material in the soil and on vegetable matter pulled down into the burrow at night. Soil which has passed through its body is voided from the anus as a worm cast. Its burrowing improves the drainage and aeration of soil, while the vegetation it pulls into its burrow adds humus. It is hermaphrodite and produces both male and female gametes at the same time. A complicated behaviour pattern has evolved allowing exchange of sperms without the risk of self-fertilisation. The eggs are laid in a cocoon and develop directly into young worms without a larval stage.

Figure 37 (opposite, below) *Hirudo* is a leech found in ponds, streams and wet soil, where it feeds as an ectoparasite on the blood of vertebrates. After feeding it drops off, its three sharp teeth leaving a Y-shaped wound from which blood flows readily because of the injection of *hirudin* an anticoagulant from the leech's salivary glands. A large blood meal can be stored in distensible gut caeca and feeds the leech for several months. *Hirudo* can swim by undulating its body and also loop using its suckers.

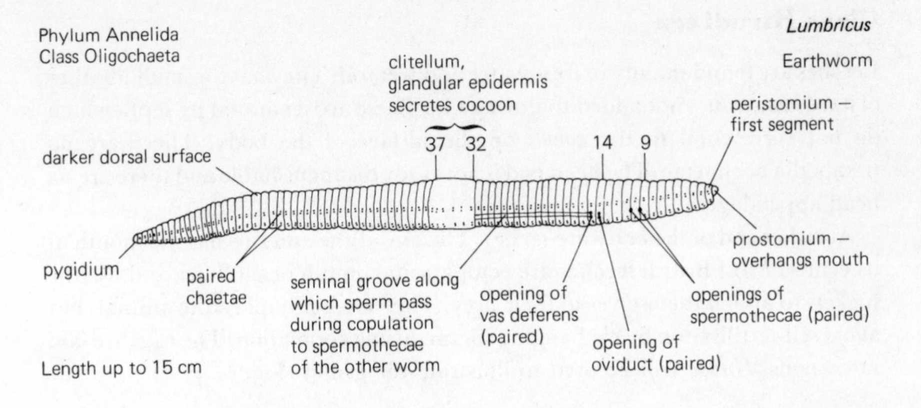

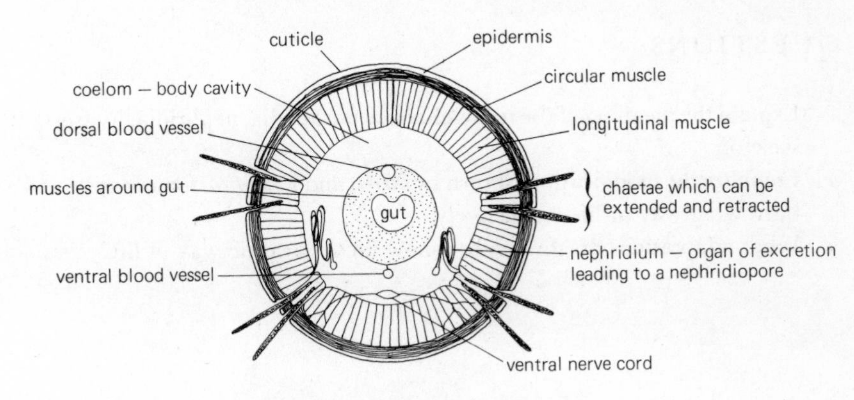

Diagrammatic T.S. of earthworm

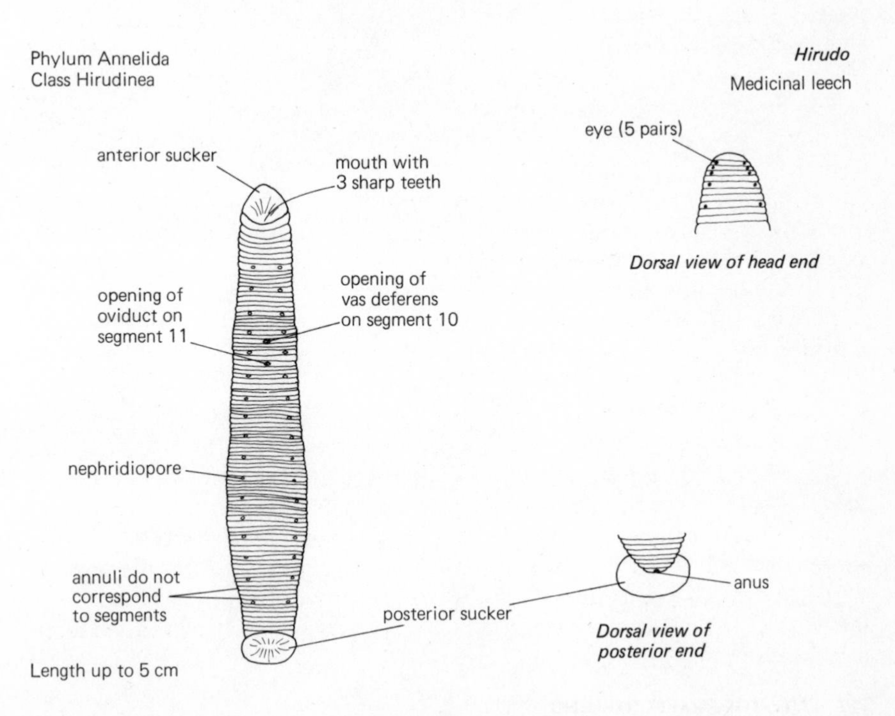

Dorsal view of head end

Dorsal view of posterior end

Class Hirudinea

Leeches are found mainly in freshwater and wet soil. They have a small number of fixed segments – not added to during life. These are separated by septa which do not correspond to the *annuli* on the surface of the body. There are no parapodia or chaetae. The head end is not easily distinguishable and there are no head appendages.

At either end of the body are *suckers*. The one at the anterior has the mouth at its centre. Most British leeches are ectoparasites on fish or molluscs and use the suckers to attach themselves to their prey. They are hermaphrodite animals but avoid self-fertilisation by exchanging sperm during copulation. The eggs are laid in cocoons. *Hirudo* will be used to illustrate the group (Figure 37).

QUESTIONS

1. Explain the meaning of the terms chaetae, parapodia, nephridia, hydrostatic skeleton.
2. Compare the modifications shown by *Nereis* and *Lumbricus* for an active and a burrowing way of life respectively.
3. What adaptations do the leeches show for a parasitic way of life?

10 Phylum Arthropoda

The success of any plant or animal group is usually measured by the number of individuals and the number of species in the group. It should, however, be remembered that both these criteria are partly determined by the size of the organisms – the smaller they are, the more the environment can support and the more habitats will be available to them. The number of species is possibly more meaningful as it also tells us something about the adaptability of the group and reflects the success with which they have coped with the challenge of new environments. Judged by both these criteria the arthropods are undoubtedly the most successful group in the animal kingdom. They comprise more than three-quarters of all known animal species and one class alone, the Insecta, includes more species than all other groups put together. It is therefore not surprising that arthropods abound in almost every type of habitat.

Arthropods show clear affinities with annelids and are thought to have evolved from some annelid-like ancestors. One small group of arthropods, the onycophora or velvet worms show both annelid and arthropod characteristics and are thought to have diverged from the main lines of arthropod evolution at a very early stage.

GENERAL FEATURES

Arthropods are bilaterally symmetrical, metamerically segmented, triploblastic coelomates. The coelom is much reduced compared with that of annelids, being restricted to small cavities in the excretory and reproductive systems. The cavity surrounding the organs is a blood-filled haemocoel not a true coelom. The circulation differs from that of annelids in that it is an *open circulatory system* – the blood is not contained in vessels throughout the whole system but instead flows from vessels into the haemocoel.

The name of this phylum refers to the jointed legs which are a characteristic feature of this group. In primitive arthropods all the segments bear one pair of jointed appendages but with the increased specialisation of body regions, that has developed during the evolution of this group, there has been a tendency towards a reduction in the number of appendages. In most forms the body is differentiated into three regions, head, thorax and abdomen. The head is always well developed and made up of six segments, although fusion of segments means this is rarely apparent.

Another important characteristic of arthropods is the presence of a rigid body covering, the *cuticle*, made of chitin and protein and often strengthened with other materials (Figure 38). This rigid structure provides support for the body. In addition, inward projections of the cuticle provide points for muscle attachment. Thus the arthropod cuticle fulfils much the same functions as the vertebrate skeleton and is accordingly called an external or *exoskeleton*. Joints between segments and between the plates which make up the exoskeleton covering each segment (Figure 39) allow movement. These joints or *articular membranes* are composed of a layer of thin flexible cuticle lacking an exodermis.

The special properties of the exoskeleton are thought to be largely responsible for the success of this group. Chitin is a very adaptable material. This has made possible the many modifications of the appendages which have enabled these animals to fill so many different ecological niches. The exoskeleton is also very

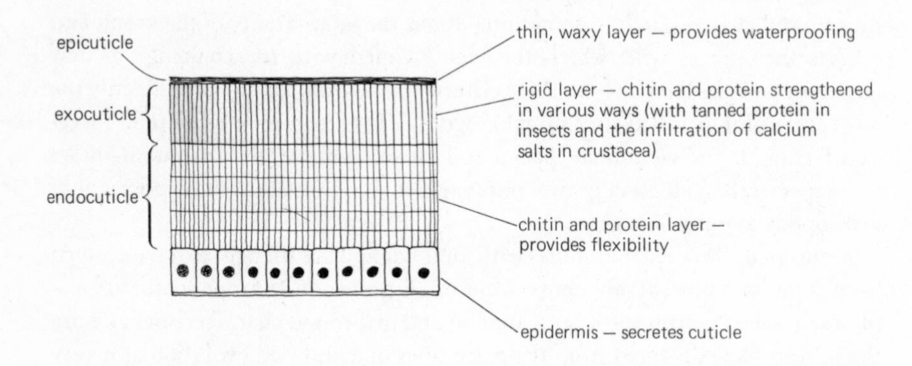

Figure 38 Diagram of transverse section through generalised arthropod cuticle

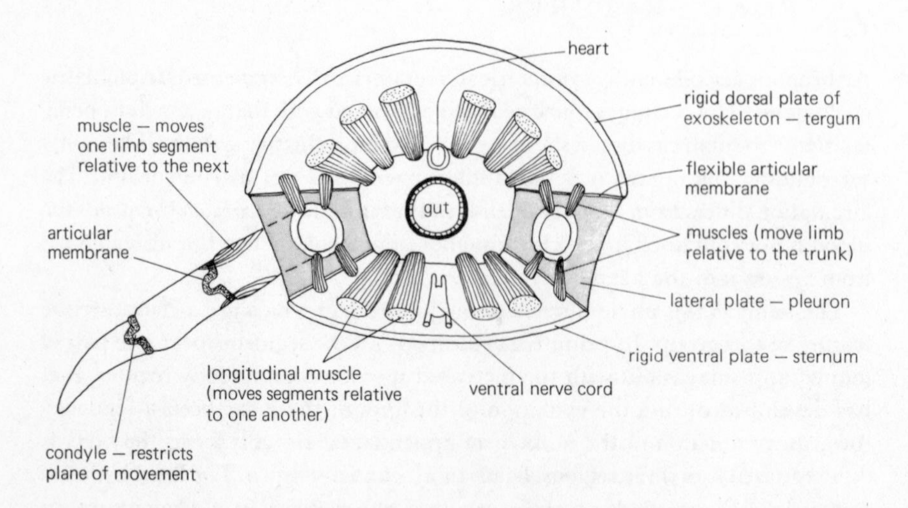

Figure 39 Diagrammatic three-dimensional section through a generalised arthropod segment

impermeable. This reduces problems of water loss and osmoregulation and was undoubtedly an important factor in their successful colonisation of the land.

On the other hand, the rigidity of the cuticle does pose problems, the most important one being that, once the cuticle has hardened, growth is severely restricted – a situation that probably accounts for the generally small size of arthropods. This problem has been partly overcome by the periodic moulting of the cuticle, known as *ecdysis*, which allows the growth of the animal while the new cuticle is hardening. The stages between moults are known as *instars*.

Most terrestrial arthropods have a system of air-filled tubes or *tracheae* for gas exchange. These open to the outside through a series of pores or *spiracles*. Most aquatic forms respire through gills, which are thin, cuticular extensions of the body wall.

CLASSIFICATION

Within such a vast group it is hardly surprising that several distinct lines of evolution can be distinguished and the group is accordingly subdivided into the following classes:

1. CRUSTACEA (crabs, prawns, lobsters, woodlice and barnacles)
2. MYRIAPODA (centipedes and millipedes)
3. ARACHNIDA (spiders, harvesters, ticks, mites, pseudo-scorpions and scorpions)
4. INSECTA
5. MEROSTOMATA (king crabs and horseshoe crabs) This group was much larger in Paleozoic times, 570 to 225 million years ago. There are only 5 living species.

Class Crustacea

With the exception of woodlice and certain crabs, members of this class are entirely aquatic, breathing through gills. They range in size from the microscopic to the Japanese spider crab which, with its four metre leg span, is the largest living arthropod.

The body is divided into head, thorax and abdomen, although the head and thorax are often fused forming a *cephalothorax*. The head always bears two pairs of antennae.

The exoskeleton varies in structure from the delicate, transparent shell of water fleas (Figure 40) to the tough, heavily calcified armour of lobsters and crabs. The exoskeleton frequently bears hair or spine-like projections called *setae*.

The sexes are usually separate and the first larval stage is typically an oval, unsegmented *nauplius* larva with three pairs of appendages.

There are over 20000 species of crustacea and so to give a better idea of the diversity within the group a few of the largest sub-classes are briefly described below and at least one representative from each is illustrated on the following pages.

Sub-class Branchiopoda

This group contains some 1200 species and is generally considered to be the most primitive, on account of the lack of specialisation shown in the limbs. Branchiopod means 'gill feet' and all members of this group have at least four pairs of feathery limbs which serve for gas exchange and filter feeding and in some forms as a means of propulsion. In many species a shell or carapace encloses part but not all of the body.
Examples: *Chirocephalus* (the fairy shrimp), *Artemia* (the brine shrimp) and the water flea *Daphnia* (Figure 40).

Sub-class Copepoda

There are over 4500 species of copepod. All are minute and they differ from members of the previous group in that there is no carapace and there are no limbs on the abdomen. Copepods abound in freshwater and in the sea. They form an important part of marine plankton and as such are an important food of many marine animals.
Example: *Cyclops*, a freshwater form (Figure 41).

Sub-class Cirripedia (barnacles)

Most barnacles are sessile, living attached to rocks, driftwood, ships' bottoms (where they can cause a serious fouling problem) and even the bodies of other animals, notably whales. Most are attached directly to the substrate but some, like the goose barnacles, are stalked. Their bodies are protected by a set of chalky plates and they all have feather-like limbs or *cirri* which filter out plankton from the water. Until the discovery of their nauplius larvae, barnacles were classified with the molluscs.
Example: *Balanus balanoides*, the acorn barnacle (Figure 42).

Sub-class Malacostraca

This group contains almost three-quarters of all known species of crustaceans and includes all the large obvious forms, such as lobsters, crabs, prawns and woodlice. The body is divided into a trunk, composed of eight segments, and an abdomen composed of six segments.
Examples: *Oniscus*, the woodlouse (Figure 43) and *Carcinus maenas*, the common shore crab (Figure 44).

Class Myriapoda

Myriapods are all terrestrial. They are characterised by their large number of legs. The body is divided into two regions, head and trunk. The head bears one pair of antennae and at least two pairs of mouthparts. Eyes if present are simple. The trunk is long and slender, typically with one pair of legs on each segment.

There are two main sub-classes, the Chilopoda (centipedes) and the Diplopoda (millipedes); in some classification schemes these are considered as separate classes. The differences between the two groups are illustrated in the drawings of *Lithobius* and *Iulus* (Figures 45 and 46).

Phylum Arthropoda
Class Crustacea
Sub-class Branchiopoda

Daphnia

Water flea

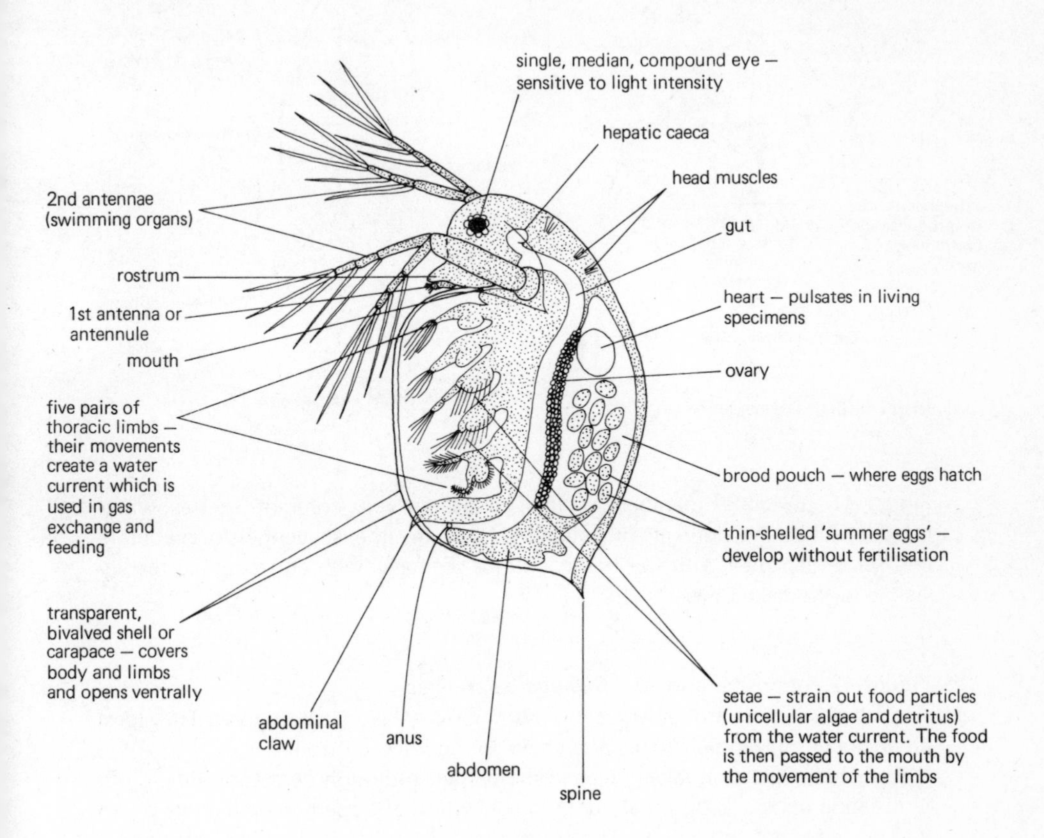

Figure 40 *Daphnia*

Water fleas are found in most types of still freshwater. The movement of the antennae produces a jerky, hopping motion (hence the name 'fleas'). Reproduction is very prolific and parthenogenesis, the development of unfertilised eggs or 'virgin birth' is common, especially in the summer months. The parthenogenetically produced 'summer eggs' hatch quickly and the young remain for some time in the brood pouch before emerging. Young produced in this way are all female. Males are produced only at certain times of the year, generally in response to the onset of unfavourable conditions such as drought, cold or food shortage. The female then produces special thick-shelled eggs with large yolk supplies ('winter eggs') which are fertilised and shed inside the brood pouch when the female moults. These eggs can survive adverse conditions, hatching when favourable conditions return.

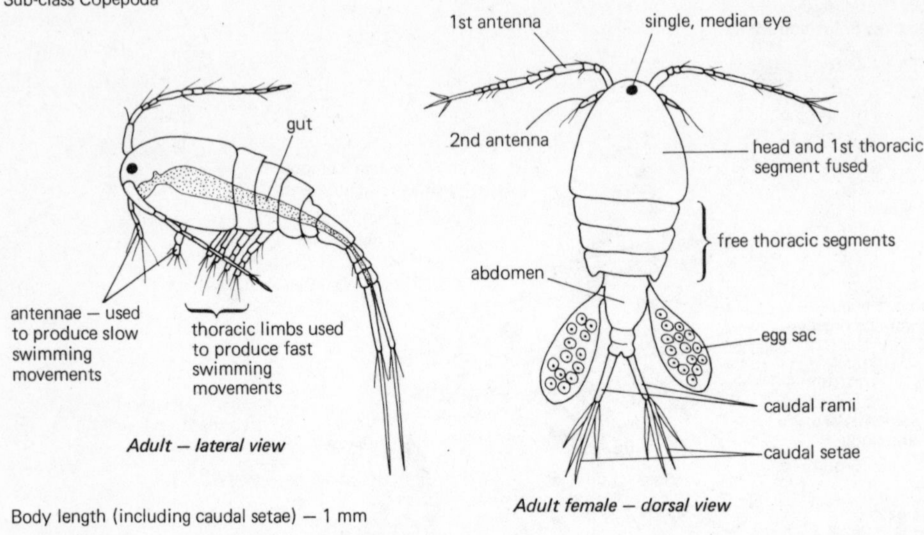

Figure 41 *Cyclops* – members of this genus are very common in freshwater ponds, ditches and slow-moving streams. They are either scavengers or predators and seize their prey with the thoracic limbs. Females with egg sacs are readily visible to the naked eye.

Figure 42 (opposite, above) *Balanus balanoides*
Acorn barnacles are probably the most common animal on rocky shores. They form an encrusting layer over rocks and shells in the intertidal zone.

Sexual reproduction takes place in the winter. Although hermaphrodite, cross fertilisation occurs, brought about by the insertion of a sperm transference organ into an adjacent individual. The eggs hatch into free-swimming, planktonic nauplius larvae; these moult giving rise to cypris larvae. The latter settle on a suitable substrate and attach themselves using special cement glands on the antennules. The adult form is assumed after a further moult.

Figure 43 (opposite, below) *Oniscus*
Woodlice are very common throughout the world. They tend to congregate under stones or in other moist situations during the day, as they are very susceptible to desiccation. They feed mainly on decaying organic matter. Breeding occurs during the warmer months. The eggs and young are carried in a special brood pouch beneath the body of the female.

Most woodlice have well developed stink glands along the sides of the body which discharge an evil-smelling substance that apparently repels predators. Some species, including *Armadillidium vulgare* – the common pill woodlouse, can curl up into a ball. This is thought to deter predators and may also be a means of reducing water loss.

Phylum Arthropoda
Class Crustacea
Sub-class Cirripedia

Balanus balanoides

Acorn barnacle

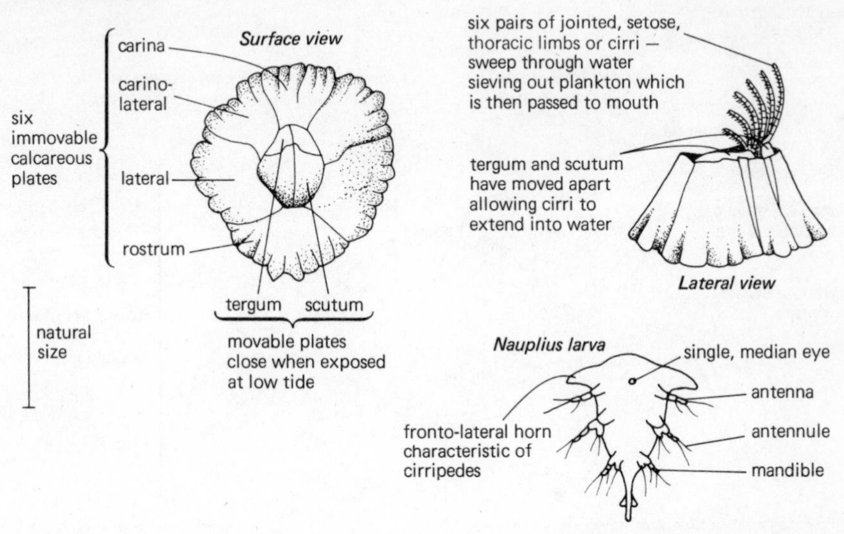

Phylum Arthropoda
Class Crustacea
Sub-class Malacostraca
Order Isopoda

Oniscus

Woodlouse

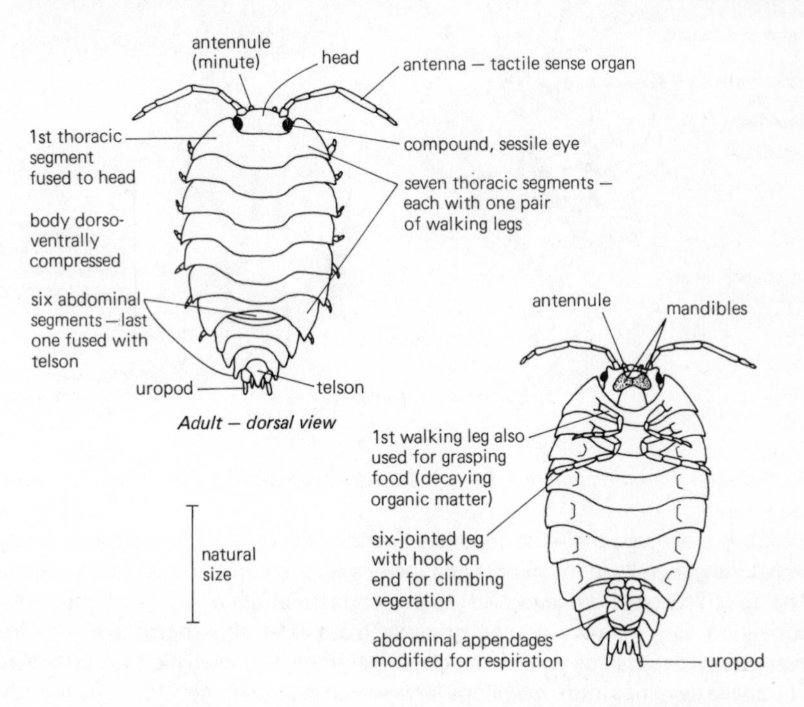

Ventral view – with all but 2 pairs of walking legs removed

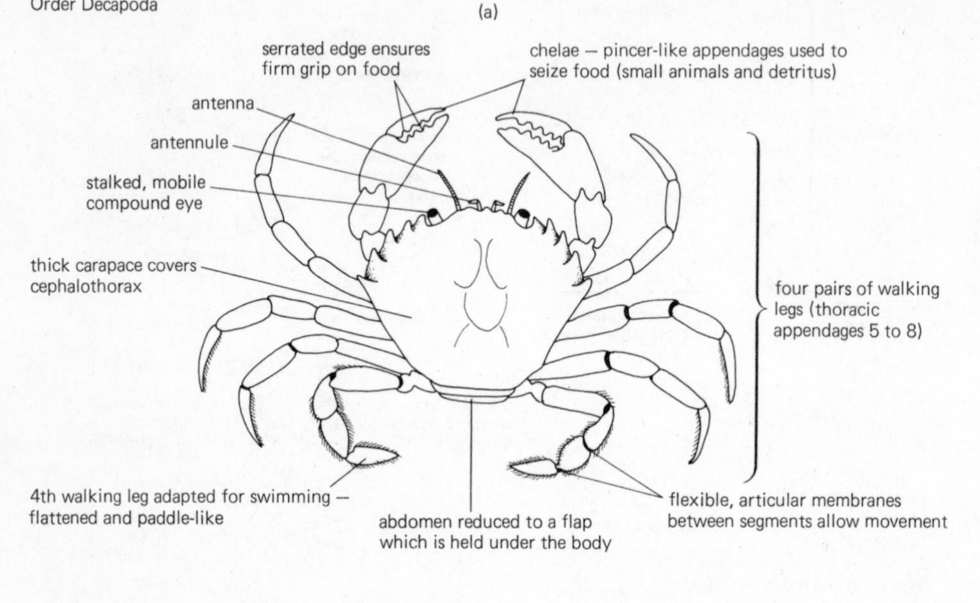

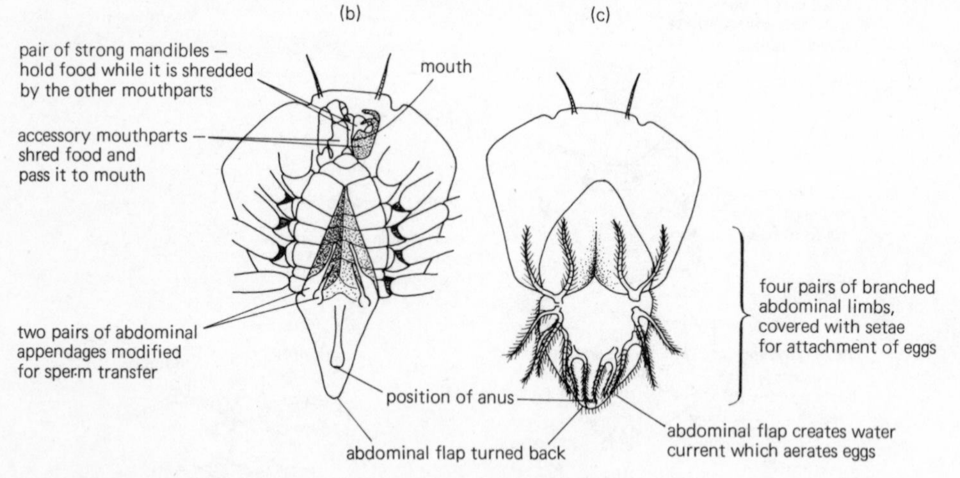

Figure 44 *Carcinus maenas*, (a) dorsal view, (b) ventral view male, (c) ventral view female

Carcinus maenas is the most common British crab. They are found on all types of seashore, generally in the middle and lower shore regions. The sexes are separate. The fertilised eggs are attached to the abdominal appendages of the female, appearing as a yellowy-orange granular mass. The eggs hatch out into free-swimming zoaea larvae with the single dorsal spine characteristic of all crab larvae. The zoaea develops into a megalopa larva which eventually develops into the adult. The egg mass should not be confused with the smooth biscuit-coloured mass of the parasitic crustacean, *Sacculina*, which often affects these crabs.

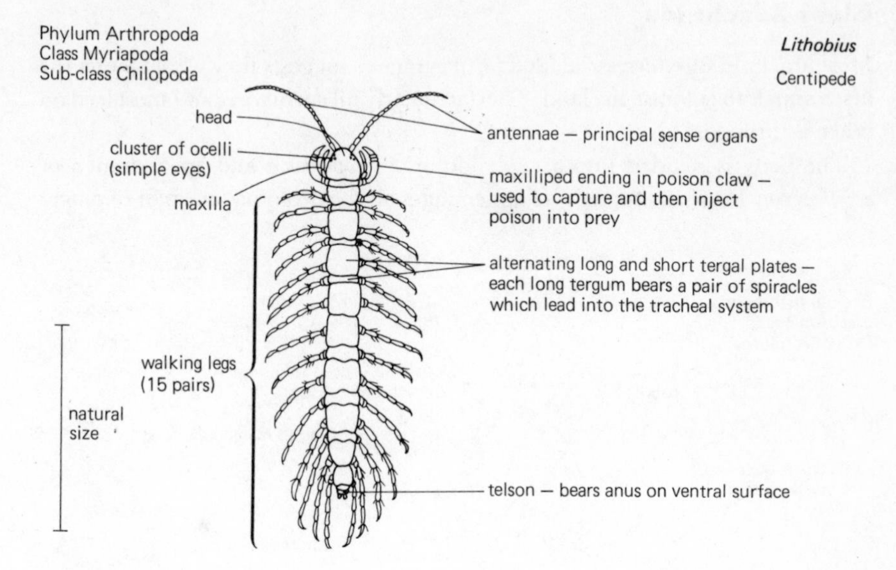

Figure 45 *Lithobius*
During the day centipedes are generally found in moist situations under bark and stones and in leaf litter. They are carnivores and hunt at night. Their prey includes slugs, earthworms and soft-bodied insects. The sexes are separate and eggs are laid singly in the soil. The young hatch as miniature adults with fewer segments, these are added during subsequent development.

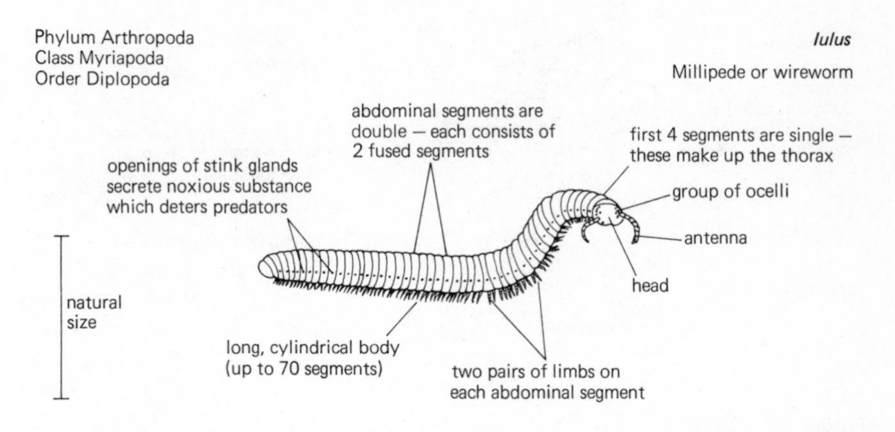

Figure 46 *Iulus*
Millipedes are also found in moist places but unlike centipedes they are vegetarian, feeding mainly on decaying plant matter and occasionally on the roots and fruits of living plants. For protection they can roll into a ball as well as using the stink glands. The sexes are separate and groups of eggs are laid in a nest in the earth prepared by the female, who watches over them until they hatch.

Class Arachnida

Most arachnids are terrestrial, and fossil evidence suggests they were among the first animals to colonise the land. They are nearly all carnivores and most feed on other arthropods.

The body is divided into a cephalothorax or *prosoma* and an abdomen or *opisthosoma*. There are six pairs of appendages on the prosoma; one pair of pincer

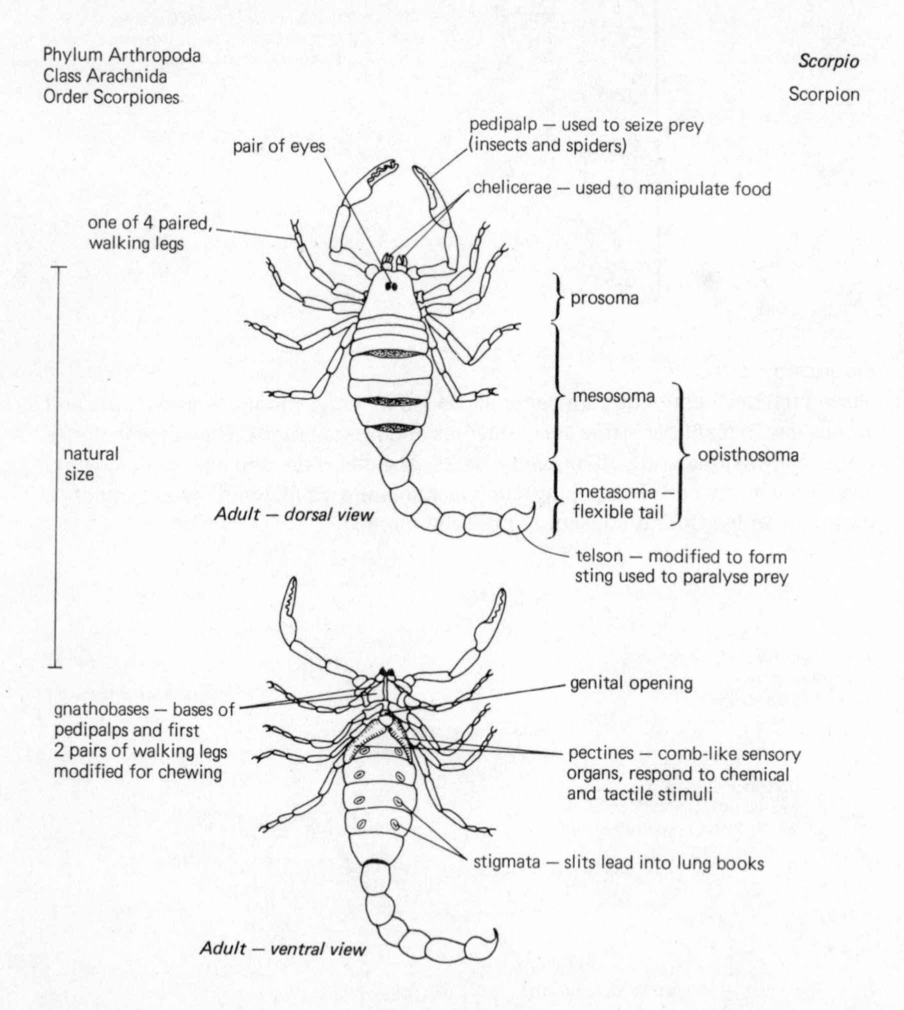

Figure 47 *Scorpio*

Scorpions are found throughout the warmer regions of the world. They are nocturnal and emerge at night from under the stones or rotten logs where they spend the day.

During a mating dance, the male places a spermatophore on the ground. This is then manoeuvred into the genital opening of the female. The young are produced viviparously. When they are born, they climb onto their mother's back where they remain for about a week before leaving to live independently.

or fang-like *chelicerae*, one pair of pincer-like or sensory *pedipalps* and four pairs of walking legs. In all arachnids either the chelicerae or the pedipalps are important offensive weapons, but never both in the same animal.

Arachnids have no jaws, food is broken up by outgrowths from the limb bases or by the chelicerae. They also lack antennae; their function is partly served by tactile hairs which cover the body and appendages.

A unique system for gas exchange has evolved in this group – the so-called *lung books*. These consist of leaf-like infoldings of thin cuticle which extend into the haemocoel. In primitive arachnids these are the only means of gas exchange but in more advanced forms they are supplemented by tracheae.

Scorpions belong to the order Scorpiones and they are generally considered to be the most primitive arachnids. They are confined to hot regions of the world and are often found in deserts although, contrary to popular belief, they are not restricted to these arid areas. Many require humid conditions and are found in tropical rain forests.

The pedipalps are large and pincer-like and are used to capture prey, mainly small arthropods. The tail or telson is modified to form a sting which injects a rapidly killing poison into the prey. Scorpions differ from other arachnids in that the young are born alive. The external features of *Scorpio* are illustrated in Figure 47.

The largest order of arachnids, the Acarinae, includes all the ticks and mites.

Mites are very widely distributed and are found in terrestrial habitats as well as in water. Most are free-living, but many of these pass through a parasitic phase in their life cycle.

Ticks are all blood-sucking ectoparasites of vertebrates. A few have very adverse effects on their hosts. Some transmit diseases such as relapsing fever and typhus, while others inject the host with a paralysing poison. *Ixodes*, the sheep tick is illustrated in Figure 48.

After the Acarinae, the Aranae or true spiders are the largest and most widely distributed group. The chelicerae are modified to form poison fangs and all have silk glands opening through spinnerets on the abdomen. The drawing of the garden spider, *Epeira* (Figure 49), illustrates the general structure of spiders. As in most arachnids, the sexes are separate and development is direct, the young hatching as miniature adults.

Class Insecta

Insects first made their appearance in the Devonian period – some 300 to 550 million years ago. They are thought to have evolved from a myriapod-like ancestor. More than five hundred thousand species of insect have been identified and it has been estimated that there may be as many as ten times that number actually in existence.

The success of this group is undoubtedly related to the evolution of wings. These confer several fairly obvious advantages – they facilitate dispersal and escape from predators, as well as other unfavourable conditions such as food shortage.

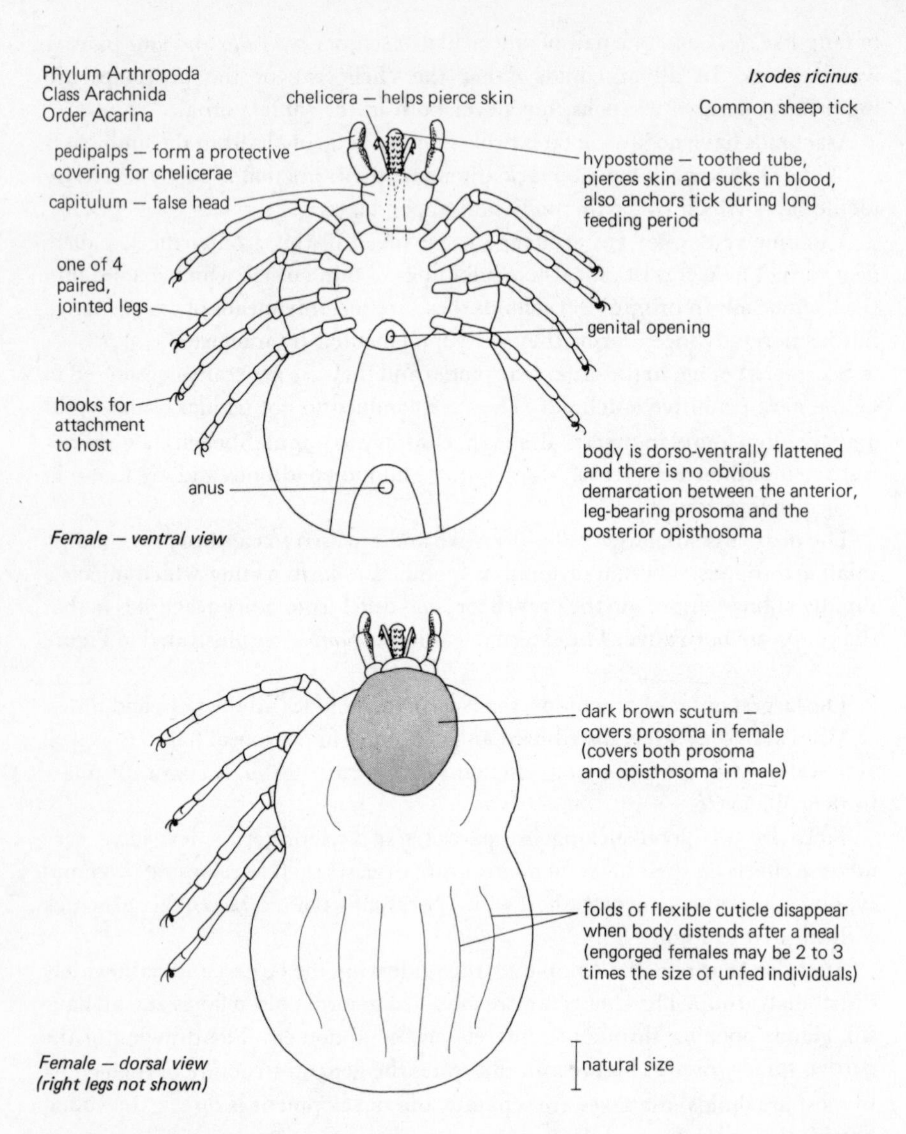

Figure 48 *Ixodes*

These ticks are most often found on sheep or cattle but they can also occur on other warm blooded animals.

Their life cycle is very complex and takes three years to complete, during which they feed on only three occasions. Eggs are laid in the soil. During the autumn the eggs hatch and a six-legged larva emerges. The following spring the larva climbs onto a host and feeds for several days, after which it falls to the ground and moults into an eight-legged nymph. The following spring the nymph climbs onto a host and again feeds for several days before falling to the ground and moulting into the adult. In the spring of the third year the adults climb onto the host for their final feed. They then drop to the ground and mate, the eggs are laid and the adults die.

These ticks are of considerable economic importance as they transmit a number of disease-causing organisms such as *Babesia* which causes redwater fever in cattle.

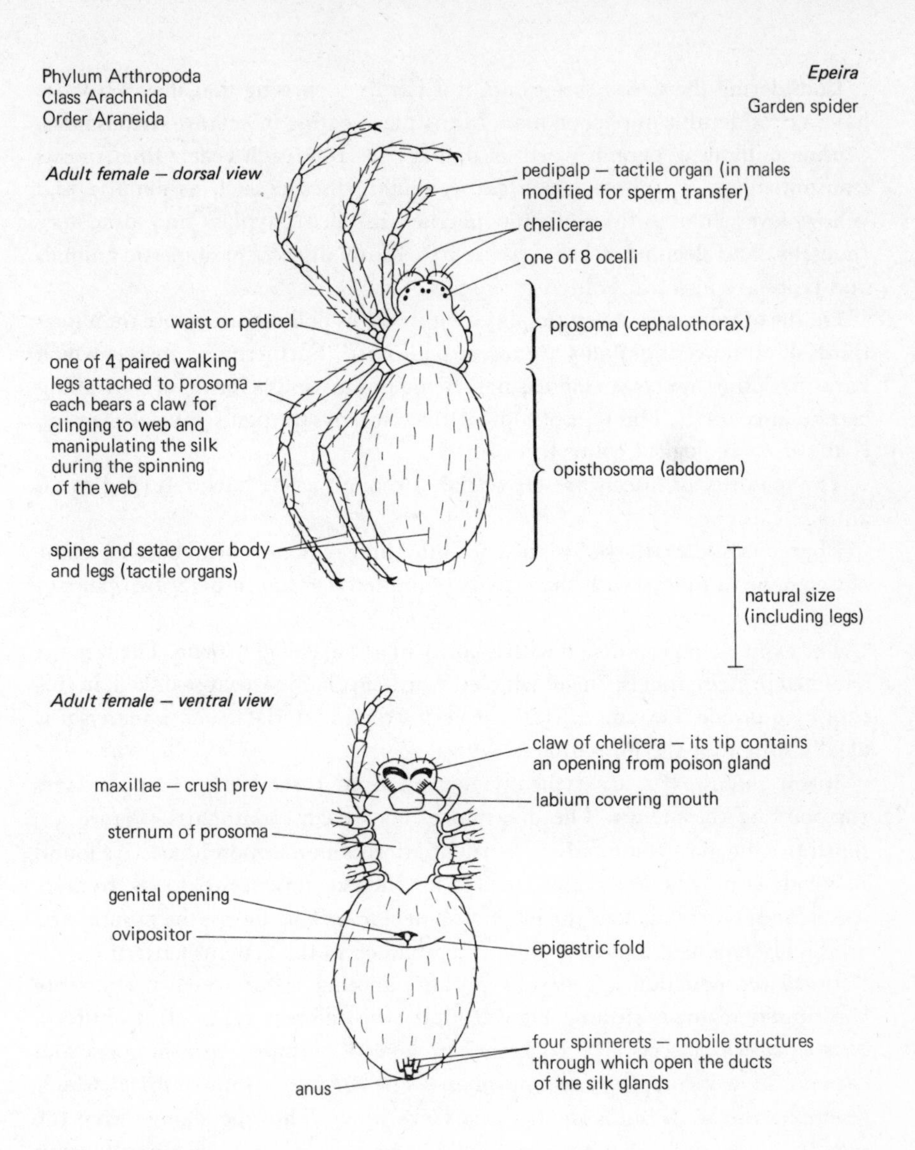

Figure 49 *Epeira*

The garden spider is a predator and feeds on small arthropods which become trapped in its web. Prey are immobilised by poison injected into them by the chelicerae and then wrapped in silk to be eaten when required. Spiders can only take in liquid food and so enzymes are poured over the food. Once these have digested the food it is sucked in by the pumping action of the pharynx and stomach.

Before looking for a mate the male constructs a silken platform on which some sperm is deposited. The sperm is then taken up by the tips of the pedipalps. On finding a mate the male has to perform a complex sequence of web-tapping to suppress the aggression of the female. During mating the male transfers the sperm into the spermothecae of the female and releases the sperm. Eggs are laid in silken cocoons. In most species, eggs are laid in the autumn and the young spiders emerge the following spring.

Considering the size of this group, it is hardly surprising that their activities have a considerable impact on man. Many plant-eating insects are serious pests, causing millions of pounds worth of damage to crops each year. Other insects transmit micro-organisms which cause human diseases such as malaria and yellow fever (mosquito), bubonic plague (rat flea), typhus and dysentery (housefly) and sleeping sickness (tsetse fly). Many diseases of domestic animals and crops are also transmitted by insects.

On the other hand, insects do play a vital role as pollinators – more than two-thirds of all flowering plants are insect pollinated. Furthermore, insects which parasitize other insects are increasingly being put to use as a means of controlling certain insect pests. This type of control, that employs the pest's natural enemies, is known as biological control.

The majority of insects are terrestrial, although a few have returned to an aquatic existence.

Characteristically the body is divided into three regions, a head bearing a pair of antennae, a thorax with three pairs of jointed legs and usually two pairs of wings, and an abdomen.

The exoskeleton is stiffened with a horny material called *sclerotin*. The latter is a tanned protein, that is, one in which the protein chains are cross-linked, in this case by quinone, forming a light but very strong material – such a material is ideally suited for the formation of wings.

Insect mouthparts are typically formed from three pairs of appendages surrounding the mouth. The diagram of cockroach mouthparts (Figure 52) illustrates the most commonly occurring pattern. Similar mouthparts are found in a wide variety of insects and are used for chewing a variety of foods. In more specialised feeders, such as the blood and nectar sucking forms, the mouthparts are highly modified and bear little resemblance to the general pattern.

Insect reproduction is generally prolific, another factor which presumably contributes to their success. They exhibit two different types of life history. *Hemimetabolous* insects like cockroaches, locusts, termites, grasshoppers and earwigs, show incomplete metamorphosis. The egg hatches into a *nymph* which resembles the adult but is smaller and lacks wings. Thus the change from the nymph to the adult is a gradual one, involving a series of moults, during which there is an increase in size and the wings develop. In contrast, in *holometabolous* insects like flies, moths, butterflies, beetles and bees, the egg develops into a *larva* which is very different from the adult. After feeding and growth the larva enters a seemingly inactive *pupal stage*. Within the pupa extensive changes occur, involving the destruction of the larval organs and the growth of adult organs from small groups of cells called *imaginal cells*. At the end of the pupal stage the adult emerges.

Classification of insects

Insects are divided into two sub-classes as follows:

1 Apterygota This group includes five orders of wingless insects in which the wingless condition is primitive. Perhaps the most familiar of these are the silverfish, often found in houses, where they feed on paper, glue, spilt flour, etc.

2 Pterygota These are either winged insects, or if wingless, the condition is secondary, that is they have evolved from winged ancestors. There are twenty orders in this sub-class. Representatives of the nine orders described below are illustrated on the following pages.

ORDER EPHEMEROPTERA (mayflies): elongate insects with net-veined wings; adults very short lived; two or three caudal filaments; antennae and mouthparts vestigial; hemimetabolous; aquatic nymph. (Figure 50.)

ORDER ODONATA (dragonflies and damsel flies): predacious insects with long, narrow net-veined wings, very large eyes, and toothed mandibles; hemimetabolous; aquatic nymph. (Figure 51.)

ORDER ORTHOPTERA (cockroaches, crickets, grasshoppers, locusts, stick insects and praying mantids): insects with long, thread-like antennae, biting mouthparts and anal cerci; tough, leathery forewings or tegmina protect broad, membranous hind-wings when at rest; hemimetabolous. (Figure 52.)

ORDER COLEOPTERA (beetles and weevils): with over 30 000 species this is the largest order of insects; forewings heavily sclerotised to form protective covers (elytra) for the membranous hindwings; chewing mouthparts, most are plant feeders or predacious on other insects; holometabolous; some species are aquatic. (Figure 53.)

ORDER TRICHOPTERA (caddis-flies): moth-like insects with two pairs of hairy, membranous wings; poorly developed chewing mouthparts; holometabolous; aquatic larvae build portable cases in which they pupate. (Figure 54.)

ORDER LEPIDOPTERA (butterflies and moths): wings, body and appendages covered with pigmented scales; mouthparts modified as coiled proboscis used to obtain nectar from flowers; holometabolous; larvae are caterpillars which usually feed on plants. (Figure 55.)

ORDER DIPTERA (true flies): all have functional forewings and reduced, knob-like hindwings (halteres); varied feeding habits reflected in varied mouthparts; holometabolous. (Figures 56 and 57.)

ORDER HYMENOPTERA (bees, wasps and ants): winged and wingless species; many show the development of the social habit, associated with this there is often the development of different castes (polymorphism). (Figures 58, 59 and 60.)

ORDER SIPHONAPTERA (fleas): small, wingless insects with laterally compressed bodies; legs adapted for jumping with very large coxae containing enlarged muscles; mouthparts are piercing and sucking, and are used to feed on the blood of mammals and birds; holometabolous. (Figure 61.)

Phylum Arthropoda
Class Insecta
Order Ephemeroptera

Ephemera
Mayfly

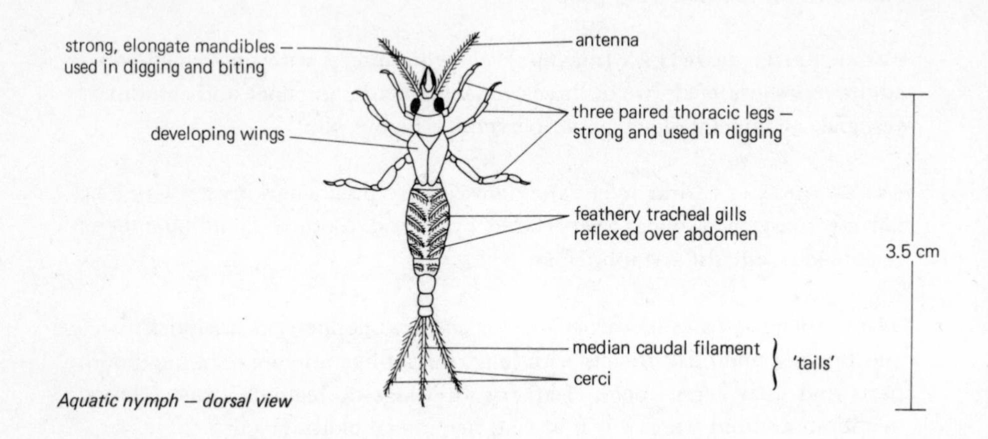

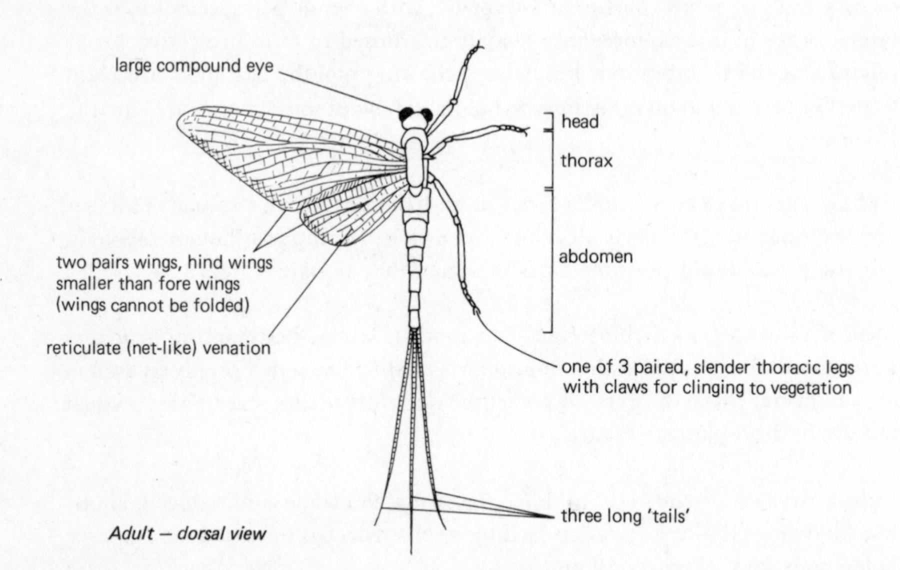

Figure 50 *Ephemera*
Adult may-flies are found only near fresh water. Their adult life is extremely short, lasting only a few hours. Adults do not feed and the mouthparts are very reduced. During their brief adult life mating occurs and the female lays thousands of eggs on the surface of the water. The newly hatched nymph burrows into the mud of the bank where it develops gills and feeds on organic matter, mainly diatoms, which it extracts from the mud it ingests. The nymph continues its growth for as long as three years, during which the wing outgrowths develop. The nymph eventually rises to the surface where it moults and fully winged forms emerge; these undergo one further moult before assuming the final adult form.

Aquatic nymph – dorsal view

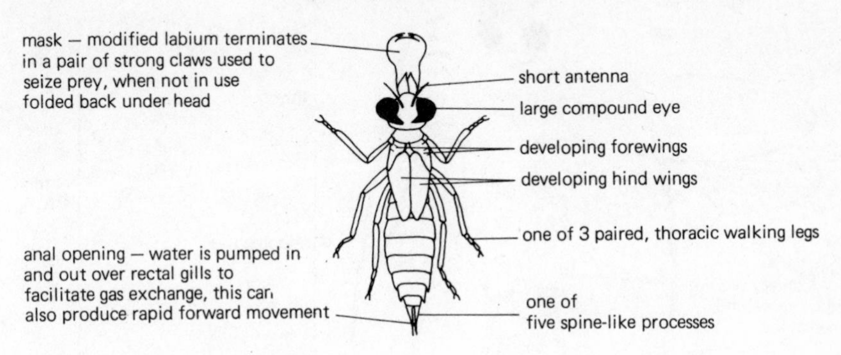

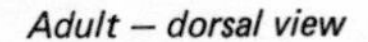

Adult – dorsal view

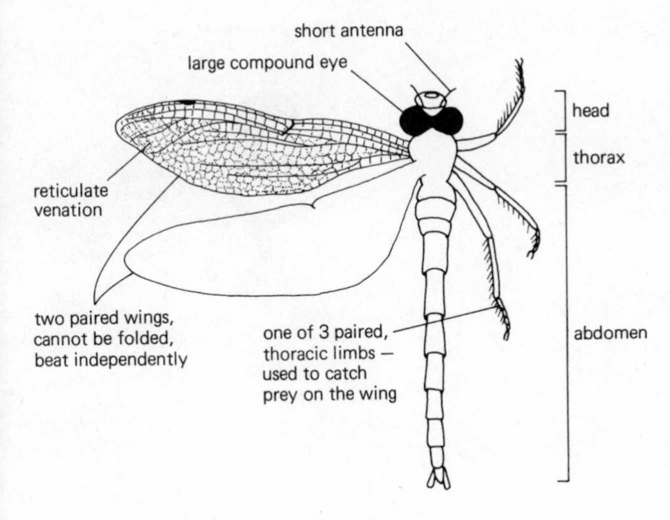

Wingspan – up to 10 cm

Figure 51 *Aeschna*
Dragonflies are carnivores feeding on other insects which they catch on the wing, using their legs. The huge eyes, which cover most of the head, contain up to 30000 lenses and enable them to detect the smallest movement anywhere near them. Mating occurs in the air and the female lays her eggs in slits in the leaves of water plants, which she makes with her ovipositor. An aquatic nymph emerges about a month later. Like the adults, the nymphs are carnivores and feed on insects and small annelids. The nymph feeds and grows for more than a year during which it moults many times. The wings develop gradually and when they are about 1 cm long the nymph climbs up a plant stem out into the air. The final moult then occurs and the adult emerges. Adults live for several months before they are killed by the cold temperatures of late autumn.

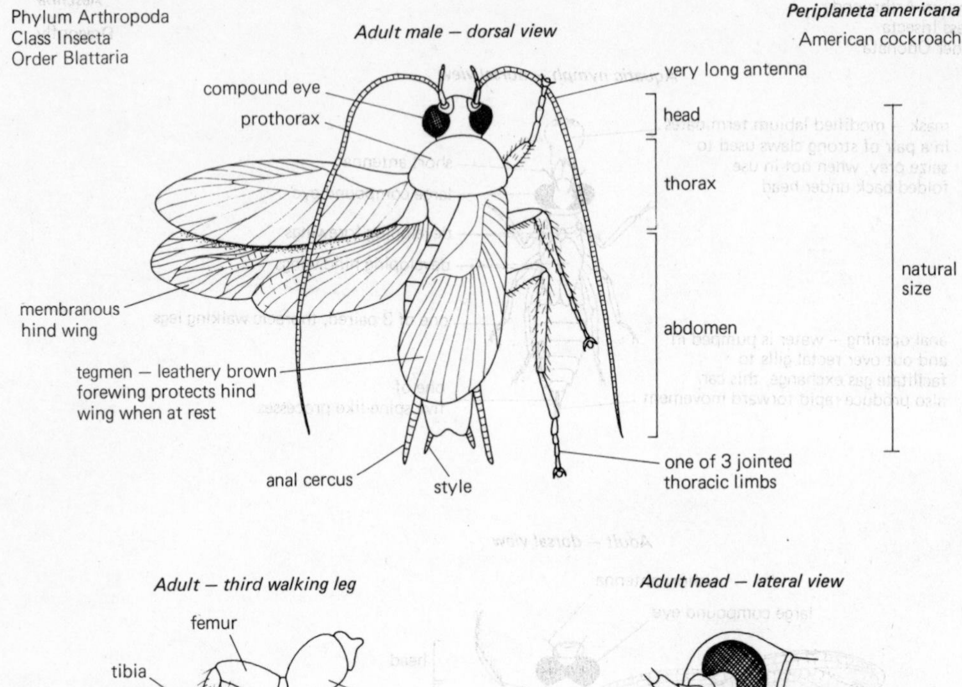

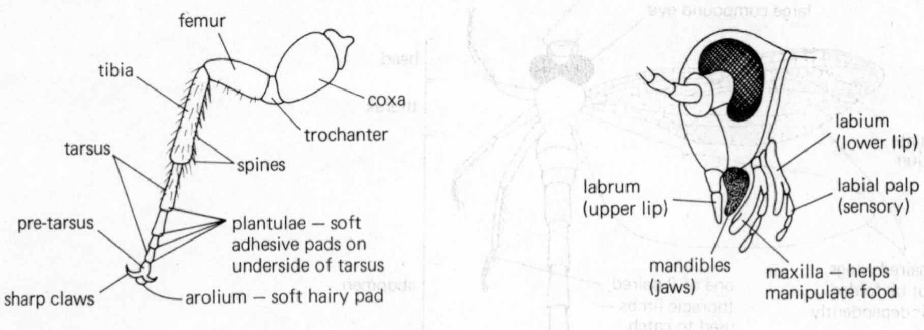

Figure 52 *Periplaneta americana* is a flattened brown insect which, as its name suggests, was brought over from America by the early trading ships. Naturally tropical or sub-tropical forms, in our colder climate they are normally found in heated buildings where they can be a pest. They live in cracks in walls and emerge at night to feed on almost any kind of organic material. They run very quickly and only rarely fly. The eggs are laid in batches of sixteen in a purse-like ootheca. The nymphs that hatch out of the eggs are small, wingless versions of the adult whose development to maturity takes five years.

Phylum Arthropoda
Class Insecta
Order Coleoptera

Gyrinus
Whirligig beetle

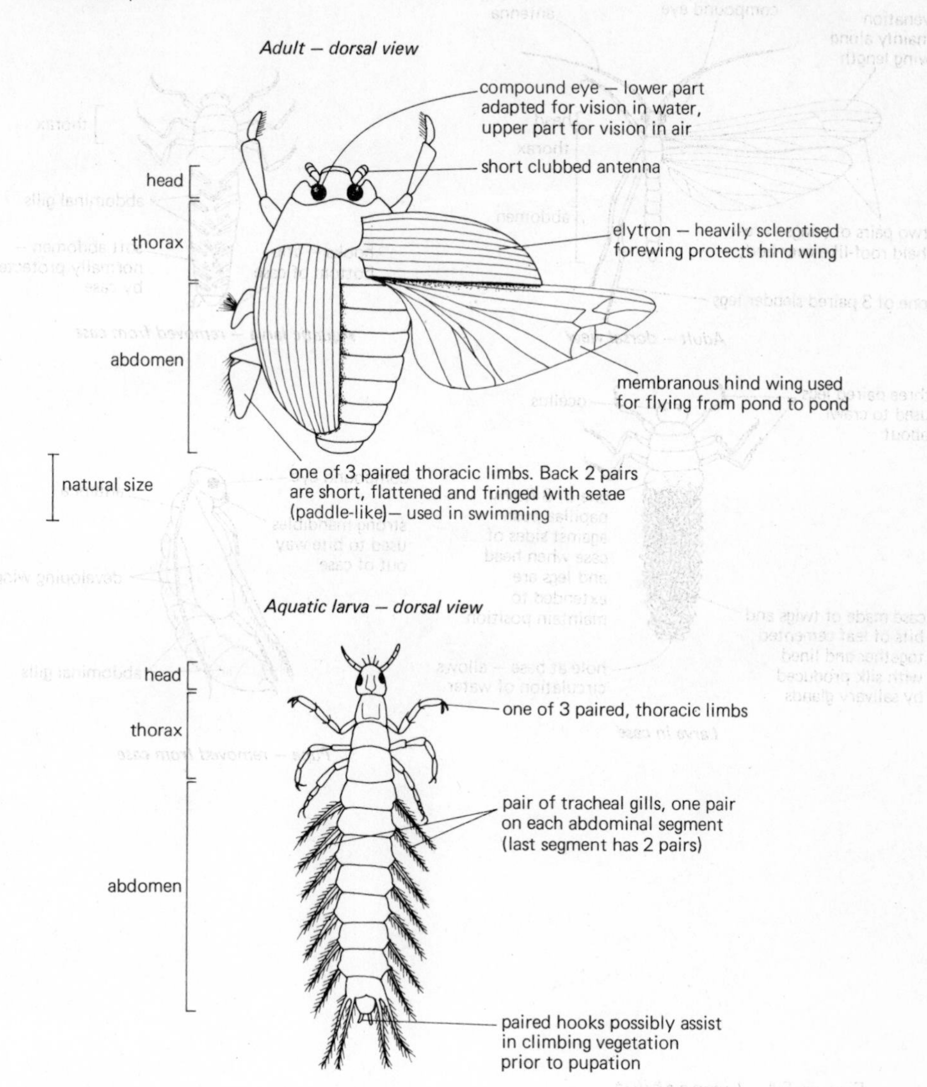

Figure 53 *Gyrinus*

These small, black beetles live on the surface of ponds and slow moving streams. They swim around in circles, seizing insects that fall onto the water. They are very sensitive to vibration and if disturbed will either dart away or dive, carrying with them an air bubble for gas exchange. Females lay their eggs on submerged vegetation in the spring. The larvae are completely aquatic and feed on insects and sometimes vegetation. At the end of July they leave the water by climbing the stems of water plants. They then spin a cocoon, inside which they pupate. A month later the adults emerge. These generally fly to other ponds before hibernating in the mud for the winter.

Phylum Arthropoda
Class Insecta
Order Trichoptera

Limnophilus

Caddis-fly

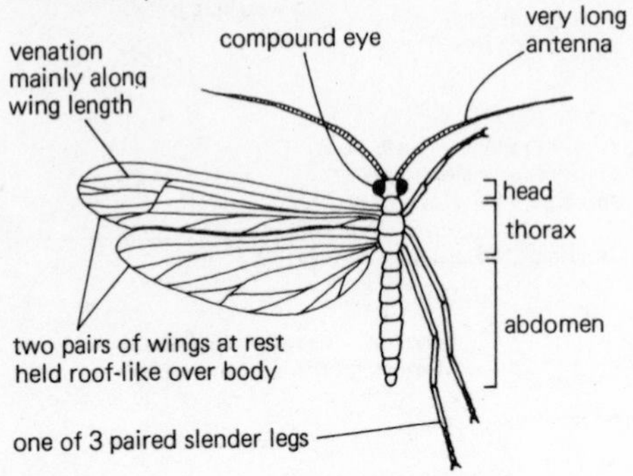
venation
mainly along
wing length
compound eye
very long
antenna
head
thorax
abdomen
two pairs of wings at rest
held roof-like over body
one of 3 paired slender legs

Adult – dorsal view

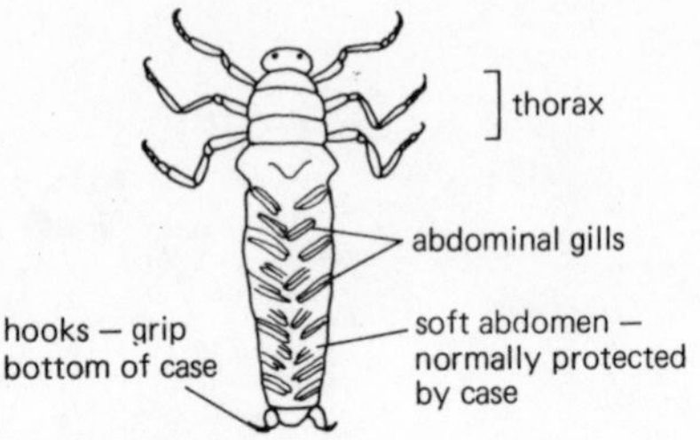
thorax
abdominal gills
hooks – grip
bottom of case
soft abdomen –
normally protected
by case

Aquatic larva – removed from case

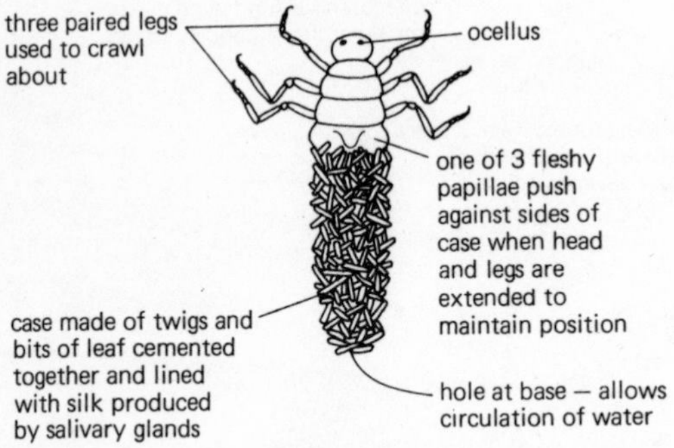
three paired legs
used to crawl
about
ocellus
one of 3 fleshy
papillae push
against sides of
case when head
and legs are
extended to
maintain position
case made of twigs and
bits of leaf cemented
together and lined
with silk produced
by salivary glands
hole at base – allows
circulation of water

Larva in case

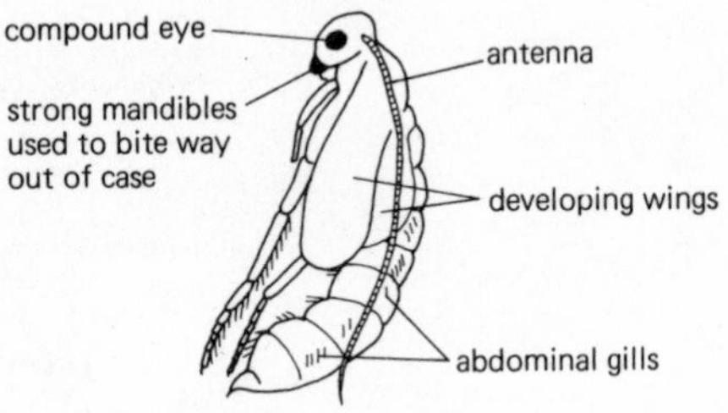
compound eye
antenna
strong mandibles
used to bite way
out of case
developing wings
abdominal gills

Pupa – removed from case

Figure 54 *Limnophilus*
Adults are found near stagnant or slow moving fresh water. They are liquid feeders and their mouthparts are much reduced. Eggs are laid by the females in the spring and summer under the surface of the water, generally attached to stones or submerged vegetation. On hatching the larva makes a protective case in which it lives. The materials of which the case is built vary from species to species and, to some extent, with the materials locally available. The cases of this genus are so heavy that they are unable to swim as some genera do, but instead crawl around. They feed mainly on decaying plant material. They pupate inside the case after they have attached it to some support and blocked the case entrance. After a few weeks or in some instances the whole winter, the pupa bites its way out of the case, swims to the surface and then sheds its pupal skin and the winged adult emerges.

Phylum Arthropoda
Class Insecta
Order Lepidoptera

Pieris brassicae

Large white butterfly

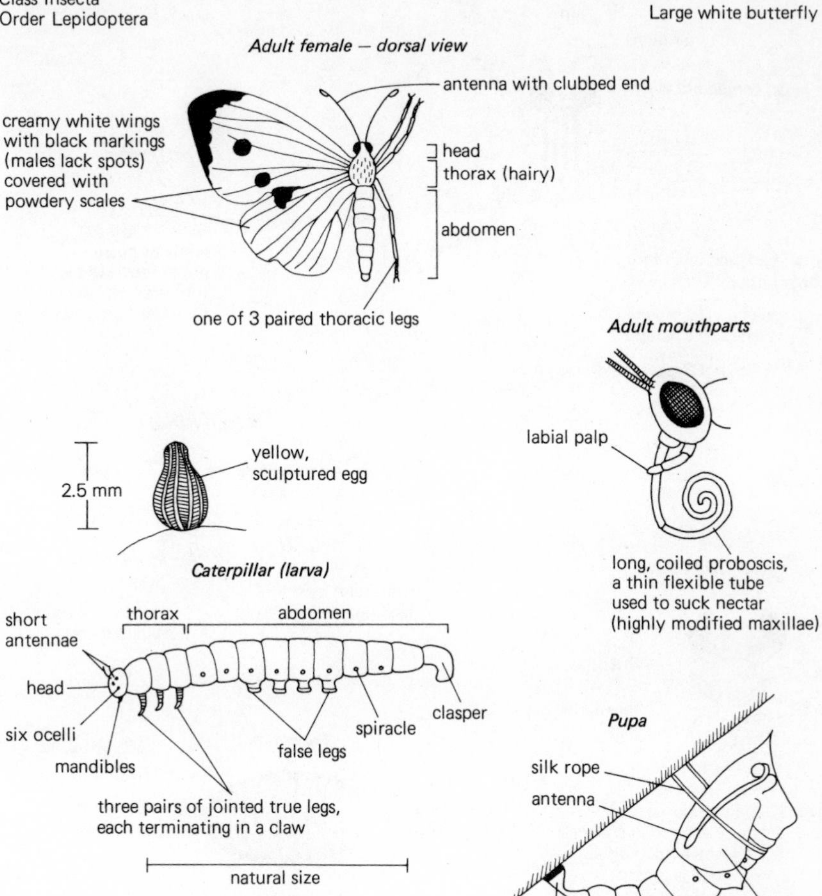

Figure 55 *Pieris brassicae*

The first adults appear in late April. They feed on nectar and live for about three weeks during which they mate. The female lays eggs on the underside of cabbage or nasturtium leaves. The eggs hatch after about eight to ten days. The caterpillar which emerges feeds on the leaves and grows very rapidly, moulting four or five times within a month. It then leaves the food plant and finds a sheltered wall or fence where it spins a silk pad to which it attaches itself by means of the clasper. It also spins a silk rope which supports the anterior end. The silk is produced by modified salivary glands which open to the surface by a spinneret behind the mouth. Once attached the caterpillar moults and the resulting pupa attaches itself to the silk pad by a cluster of hooks (cremaster). After three weeks the adults start to appear, they mate and lay eggs. The second generation of pupae generally overwinter and the adults emerge the following spring.

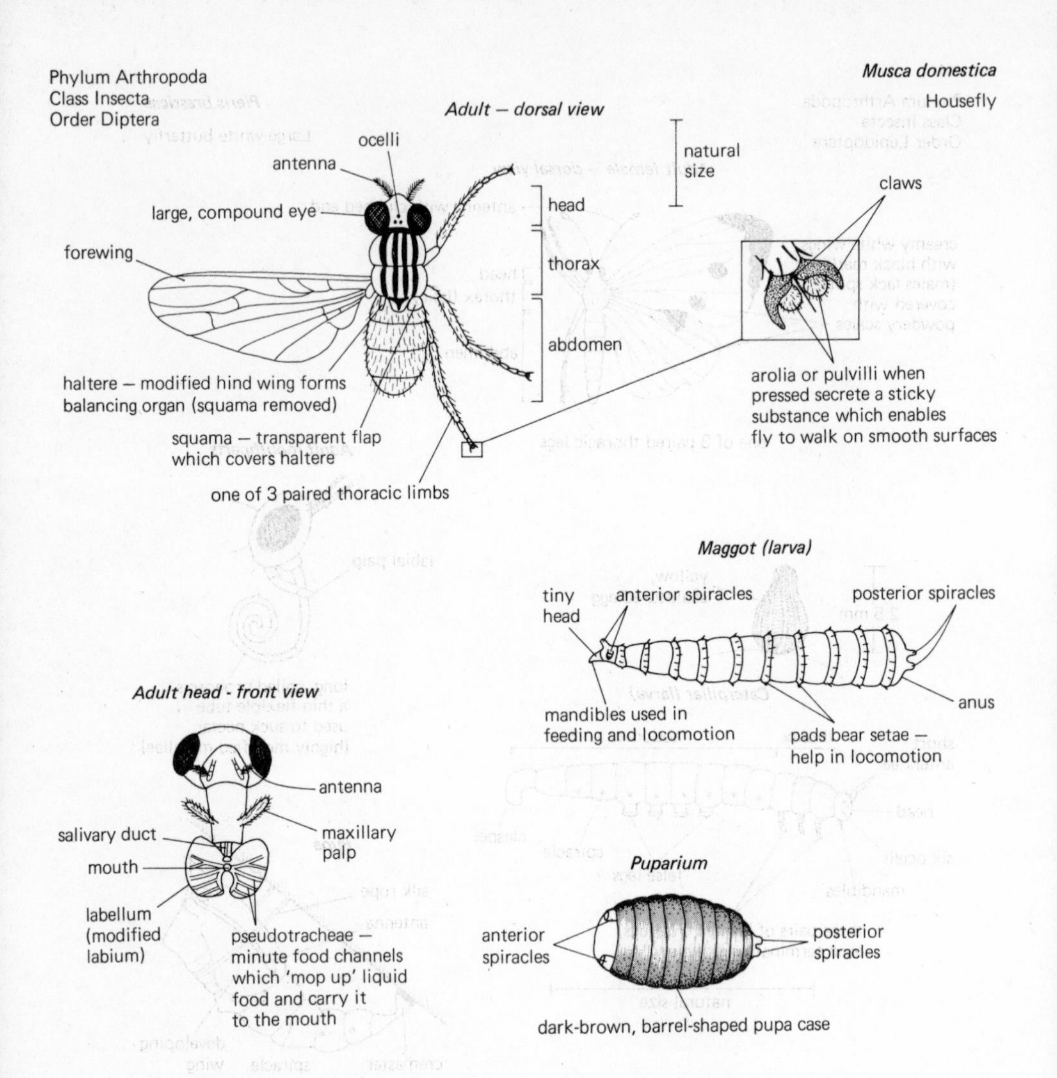

Figure 56 *Musca domestica*

Houseflies feed on almost any kind of organic matter using their peculiar sucking apparatus. The female lays her eggs in up to six batches with about 100 eggs in each on decaying organic matter during the summer months. The eggs which are only about 1 mm long hatch within twenty-four hours. The tiny white maggots (larvae) which emerge feed almost continuously on decaying organic material. They moult two or three times and then pupate. Some pupate over winter while others complete their development after only three or four days. The adults emerge by pushing open the lid of the pupa case.

These flies constitute a serious health hazard. As a result of their unclean feeding habits they can carry many disease-causing micro-organisms; these include the causative agents of typhoid, cholera and possibly smallpox and poliomyelitis.

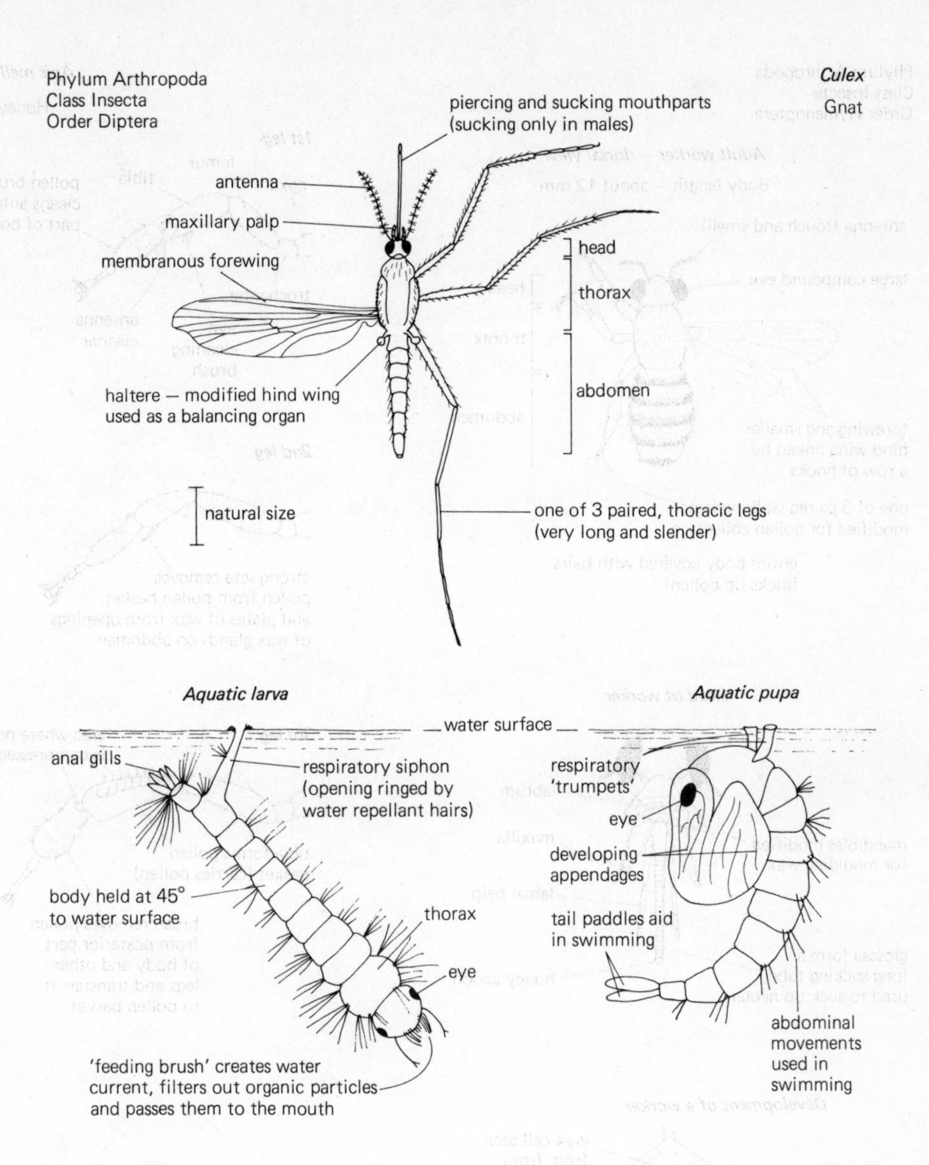

Figure 57 *Culex*

Adults feed on plant juices although females also feed on blood and in some cases are unable to lay any eggs without a blood meal. Gnats are never found far from slow-moving or stagnant water where they lay their eggs. These are laid in batches of about 200 enclosed in an adhesive envelope forming a floating raft. The eggs hatch out into free-swimming, aquatic larvae which are often to be found suspended from the water surface where they go to replenish their oxygen supplies. The larvae moult three times, at the fourth moult the pupa emerges. Although it does not feed it nevertheless moves actively about and like the larva comes to the surface to breathe. After about five days the pupal case splits and the adult emerges. Adults generally survive the winter by hibernating.

Phylum Arthropoda
Class Insecta
Order Hymenoptera

Apis mellifera

Honey bee

Adult worker – dorsal view

Body length – about 12 mm

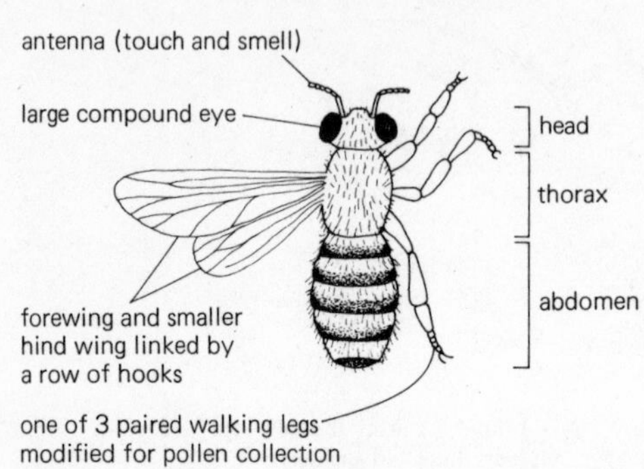

entire body covered with hairs (picks up pollen)

1st leg

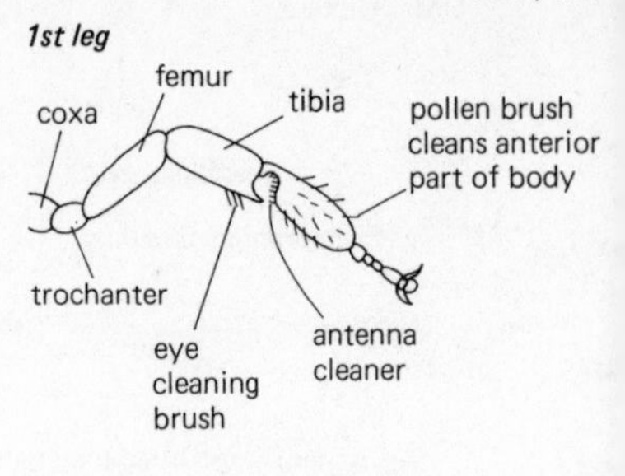

2nd leg

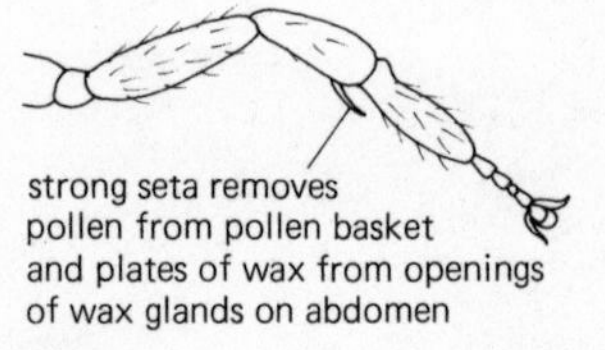

Head of worker

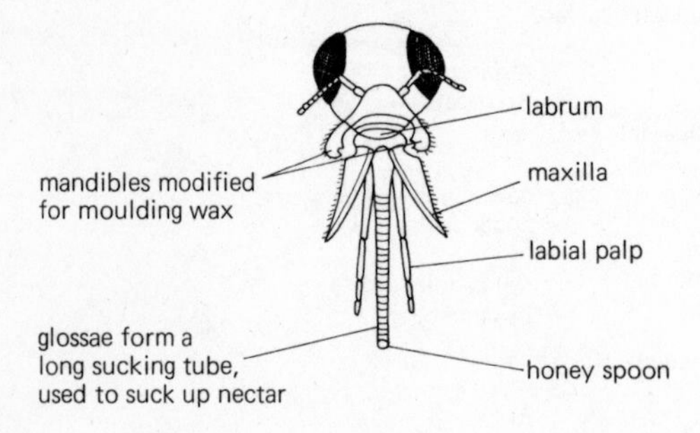

3rd leg

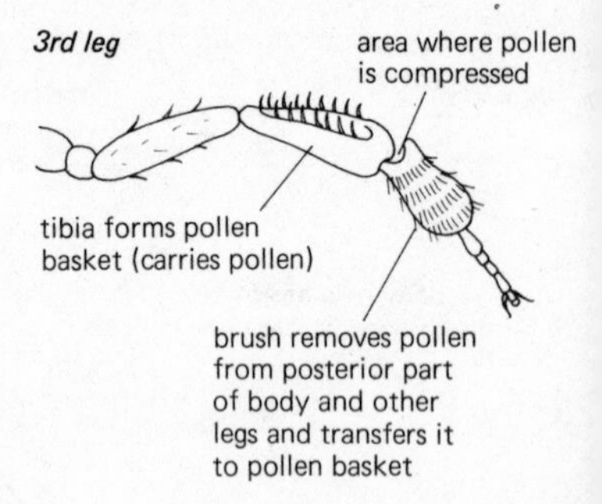

Development of a worker

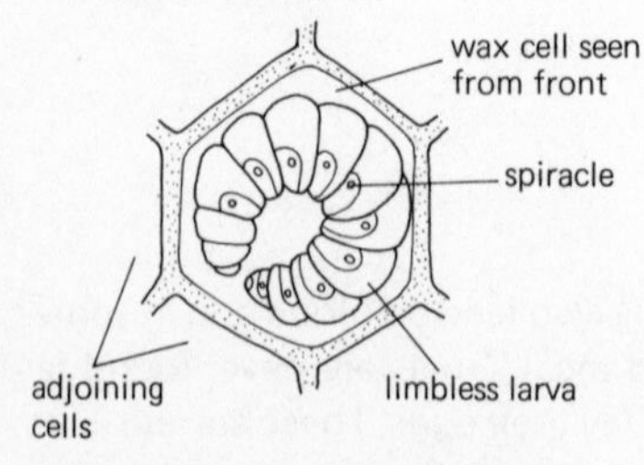

1 Eggs are laid in cells by the queen; 4 to 5 days later the larvae hatch out. They are fed by the workers and moult several times before changing into pupae

2 Side view of cell with enclosed pupa

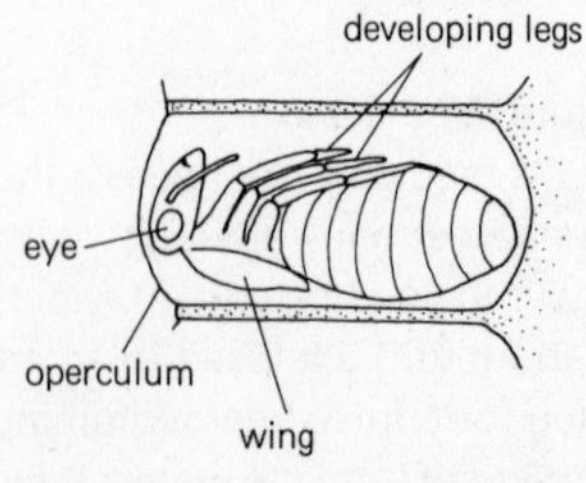

3 Twenty-one days after hatching the adult emerges from the cell

Figure 58 (opposite, above) *Apis mellifera*
Honey bees are social insects, that is they live in highly organised communities. Within a community which can number up to 50000 bees three types of individuals or castes can be distinguished. There is a single, large, fertile female, the queen, whose sole function is egg-laying which she does at the rate of two to three thousand per day. There are several hundred males or drones, these fertilise the queen during a nuptial flight after which they are turned out of the hive by the workers, the third and most numerous type.

Workers are sterile females and, as their name suggests, they perform most of the work of the hive. They only live for about eight weeks during which they perform a variety of tasks according to their age in the following sequence: looking after the young, cleaning and ventilating the hive, building wax cells (some to contain developing young and others stores of honey and pollen), guarding hive entrances and finally foraging for food outside the hive. It has been discovered that returning workers can communicate the whereabouts of food sources to other bees by means of a peculiar tail-wagging dance. They also feed other members of the hive with regurgitated food. This habit of exchanging food is probably important in maintaining the social bond. The queen produces a substance, oxodecenoic acid, which is licked from her body by attendant workers and passed around the hive during food exchanges. This 'queen substance' inhibits the development of workers' ovaries and also prevents them building any queen cells. If the hive becomes overcrowded the queen leaves the hive with about half the colony to establish a new colony. In her absence the workers build queen cells and start to rear her potential successors. The first queen to emerge usually kills all her would-be rivals by stinging them to death in their cells. She then assumes the responsibility for egg laying after her nuptial flight.

Figure 59 (opposite, below) Development of the honey bee
The caste of a bee is determined partly by its genetic make-up and partly by its diet. Drones or males are haploid and develop from unfertilised eggs while workers and queens are diploid and develop from fertilised eggs. Both female castes are fed on a protein-rich substance called 'royal jelly' for the first few days after hatching. Thereafter while the potential queens continue on this diet, worker larvae are switched over to 'bee-bread' a mixture of honey and pre-digested pollen.

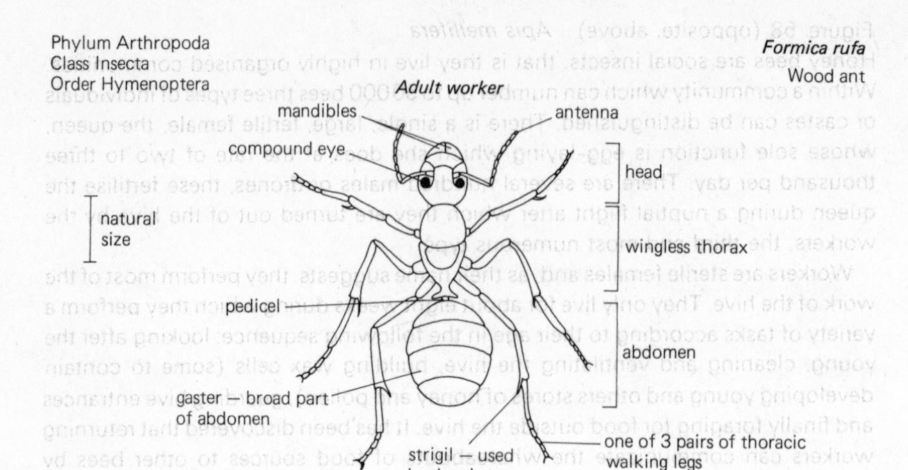

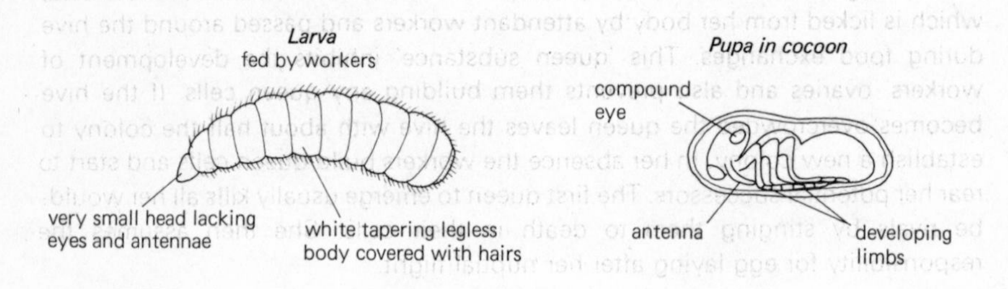

Figure 60 *Formica rufa*

Wood ant nests are a conspicuous feature of many woodland areas. They are made of wood splinters or pine needles and may be as high as 2 m. Within the nest the social organisation is very similar to that of the honey bee. The most numerous individuals are sterile female workers, who are wingless. There are several larger fertile females or queens and in the summer numerous winged males are also present. The males and fertile females are the 'flying ants' which usually swarm for a brief period prior to mating. The workers look after the hive and the young and as in honey bees their tasks vary according to their age. Young workers act as nursemaids and constantly lick the eggs and larvae to prevent desiccation and possibly to prevent bacterial or fungal growth. Older workers are employed in foraging outside the nest. They are carnivores and feed mainly on other insects. When foraging ants return to the nest they feed other ants. This feeding is thought to be important in establishing and maintaining the social bond. It probably also serves to pass around the scent peculiar to each nest which enables members of the same nest to identify one another.

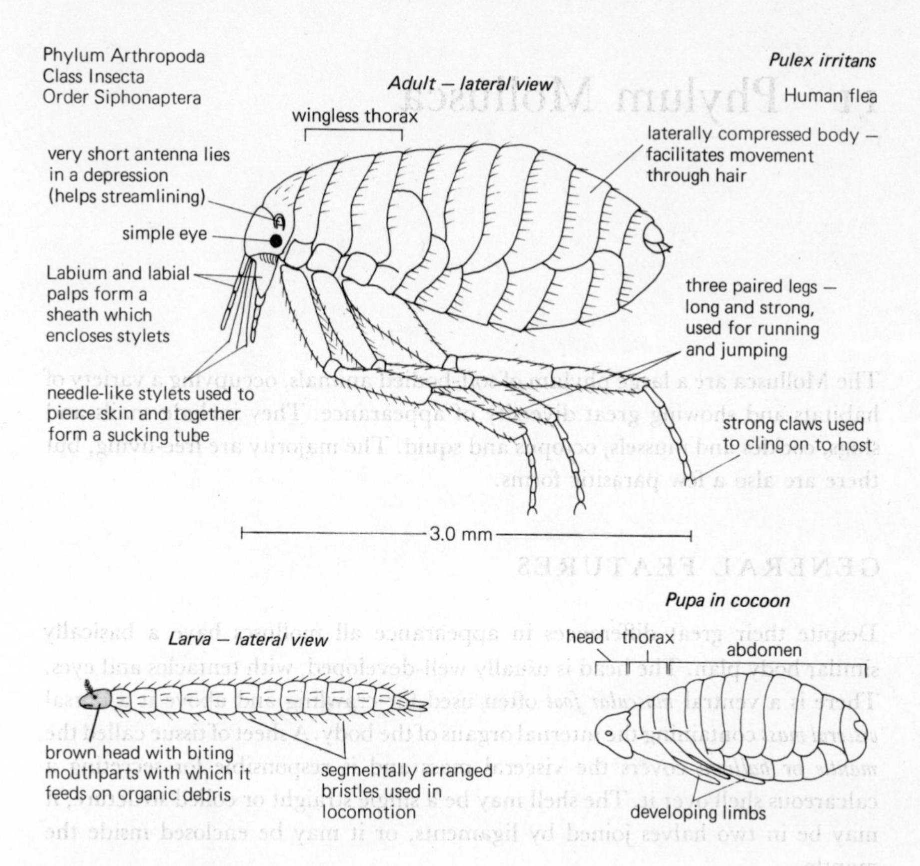

Figure 61 *Pulex irritans*
The adult flea is an ectoparasite of man and feeds on blood after first injecting saliva together with an anti-coagulant into the wound made by the piercing stylets. The female cannot produce eggs unless she has had a blood meal. After the meal a few eggs are laid between floorboards, in carpets or in the bedding of the host. The eggs are coated with a sticky substance with which they are attached to the substratum. The larva which hatches out feeds for about two weeks and normally moults twice before it makes a silken cocoon in which it pupates. Metamorphosis normally takes four to five days after which the adult emerges and seeks a host.

QUESTIONS

1 Explain the meaning of the following terms: exoskeleton, articular membrane, ecdysis and trachea.
2 Outline the features that, a) insects share with other arthropods, and b) distinguish insects from other arthropods.
3 Write an essay on the importance of insects to man.
4 Distinguish between a hemimetabolous and a holometabolous insect. Describe the life cycle of a named holometabolous insect.

11 Phylum Mollusca

The Mollusca are a large phylum of soft-bodied animals, occupying a variety of habitats and showing great diversity of appearance. They include snails and slugs, cockles and mussels, octopus and squid. The majority are free-living, but there are also a few parasitic forms.

GENERAL FEATURES

Despite their great differences in appearance all molluscs have a basically similar body plan. The head is usually well-developed, with tentacles and eyes. There is a ventral *muscular foot* often used for crawling and above it a dorsal *visceral mass* containing the internal organs of the body. A sheet of tissue called the *mantle* or *pallium* covers the visceral mass and is responsible for secreting a calcareous shell over it. The shell may be a single straight or coiled structure, it may be in two halves joined by ligaments, or it may be enclosed inside the mantle.

Inside the mantle is the mantle cavity which may contain gills (*ctenidia*) and the openings to the kidneys, reproductive system and anus. It may be possible for the foot to be drawn into the shell for protection.

The molluscs are triploblastic, coelomate animals which do not show segmentation of the body. The coelom has been reduced to small cavities around the heart, in the kidney and gonad and sometimes around the gut.

The heart circulates blood through an open blood system of haemocoels. Only the very active cephalopods have closed blood vessels. As well as its use in respiration, molluscan blood serves as a hydrostatic skeleton for extending the foot and other parts of the body.

There is a basic bilateral symmetry in the Mollusca, which has been lost in gastropods by the coiling of the body and a twisting process known as *torsion*.

Molluscs may be either hermaphrodite or of separate sexes. If external fertilisation occurs it is usually followed by two larval stages – the *trochophore* and *veliger*. Where internal fertilisation happens in cephalopods and some gastropods, yolky eggs are laid which frequently develop directly into young adults.

Many molluscs, including winkles, snails, cockles and mussels are considered good to eat, and oysters in particular are regarded as a delicacy. After a planktonic larval stage the young oysters, which are called *spat*, settle as sedentary adults. They may form extensive oyster beds, such as those at Whitstable and in the Helford River where they are cultivated commercially.

CLASSIFICATION

There are three main classes of Mollusca. The Gastropoda are the slugs, snails and limpets, where the shell is usually in one piece and coiled above a distinct foot. The Lamellibranchia are the bivalve molluscs, such as mussels where the shell is in two halves and encloses the body. The third class is the Cephalopoda, the active predacious molluscs, such as the squids which are modified for fast movement and greater awareness of their environment, through a more highly differentiated nervous system and sense organs.

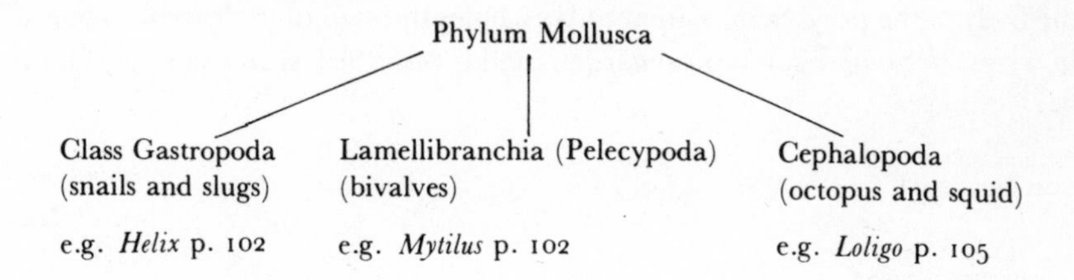

Class Gastropoda

This is the largest and most varied class of the molluscs. There are terrestrial, freshwater and marine snails, land slugs and sea slugs – the beautiful brightly-coloured nudibranchs. Their habits range from burrowing to swimming actively in the plankton, like the sea-butterflies.

During their evolution the gastropod body has developed a twist known as torsion, when the visceral mass was turned through 180° in relation to the head-foot. Before torsion the anus was at the posterior and discharged its products into the water above the gills. The gills also had to function in water and particles stirred up by the animal as it moved along. Torsion placed the anal and renal openings above the head where they discharge into the exhalant current from the gills. It also brought the gills to the anterior, just above the head, where they receive clean water and an associated sense organ, the *osphradium*, tests the water for chemicals and sediment. As well as twisting the alimentary canal, torsion also twisted the nervous system into a figure of eight.

Many gastropods also show a coiling of the visceral mass into a right-handed spiral carried above the head-foot. This was aided by the atrophy of most of the viscera on the right-hand side of the mantle cavity. As a result of torsion and coiling, the bilateral symmetry typical of other molluscs is absent from this group.

The gastropod shell coils about a central pillar – the *columella*, to which the body is attached by the *columella muscle*. When this contracts it withdraws into the shell. It re-emerges when the muscle relaxes and blood is forced into the haemocoels of the head-foot. Many aquatic snails are also able to close the shell aperture with a horny plug – the *operculum*, carried on the upper surface of the posterior end of the foot.

Adult limpets have lost the coiled shell, leaving only a single cone-shaped last whorl. Some slugs also show a reduction or even complete loss of the shell, together with uncoiling and detorsion of the body, reverting to an external bilateral symmetry.

Terrestrial gastropods are *pulmonate* – having the mantle cavity modified as a lung. Some of these have later returned to freshwater and therefore must come up to the surface to breathe.

Most gastropods are herbivores, browsing on vegetation using a *radula* – a ribbonlike structure covered with small hooks, which is scraped to and fro against a hard pad on the roof of the mouth. Others feed on living or dead animals, sometimes with a radula at the end of a long proboscis which penetrates the body of the prey. Some can even bore holes through the calcareous shells of bivalves. *Helix aspersa*, a typical garden snail is described as an example (Figure 62).

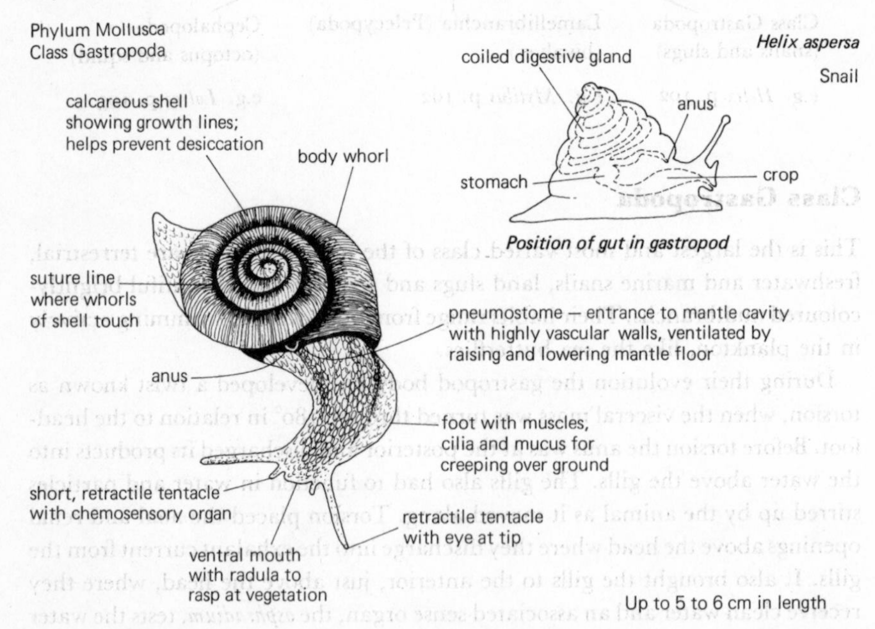

Figure 62 *Helix aspersa* is a terrestrial, pulmonate snail of gardens and farmland wherever it is relatively damp. It feeds on vegetation scraped off by the radula. In winter it hibernates, withdrawing into the shell and closing the aperture with a layer of mucus and calcium phosphate. It is hermaphrodite with a complicated process of sperm exchange during copulation so that cross-fertilisation is ensured. The fertilised eggs are deposited in damp places under soil or vegetation.

Figure 63 (opposite) *Mytilus edulis* is commonly found attached to rocks and breakwaters from mid tide level down to about 6 m. Extensive mussel beds may also build up on sandy or muddy shores if there are enough rocks for the initial settlement of larvae. They may move short distances by breaking and reforming the threads. They filter feed on detritus and plankton. Breeding occurs in spring or summer. The sexes are separate and gametes are released into the sea. After fertilisation the zygote develops first into a trochophore larva and then into a veliger, which eventually settles on the shore.

Class Lamellibranchia (Pelecypoda)

The lamellibranchs are a smaller and less varied class than the gastropods. They have become highly modified as ciliary feeders, losing the distinct head and radula, but having greatly developed the gills.

The whole body can be enclosed in the two halves of the shell hence the term *bivalve*. The two valves are held together by a *hinge* of teeth fitting into grooves and by an elastic ligament. The shell can be tightly closed by *anterior* and *posterior adductor muscles*. The ligament works in opposition to the adductor muscles and helps to open the shell.

The foot is usually laterally compressed and wedge-shaped, (Pelecypoda – hatchet foot). It may be extended and withdrawn by the *pedal muscles*. The

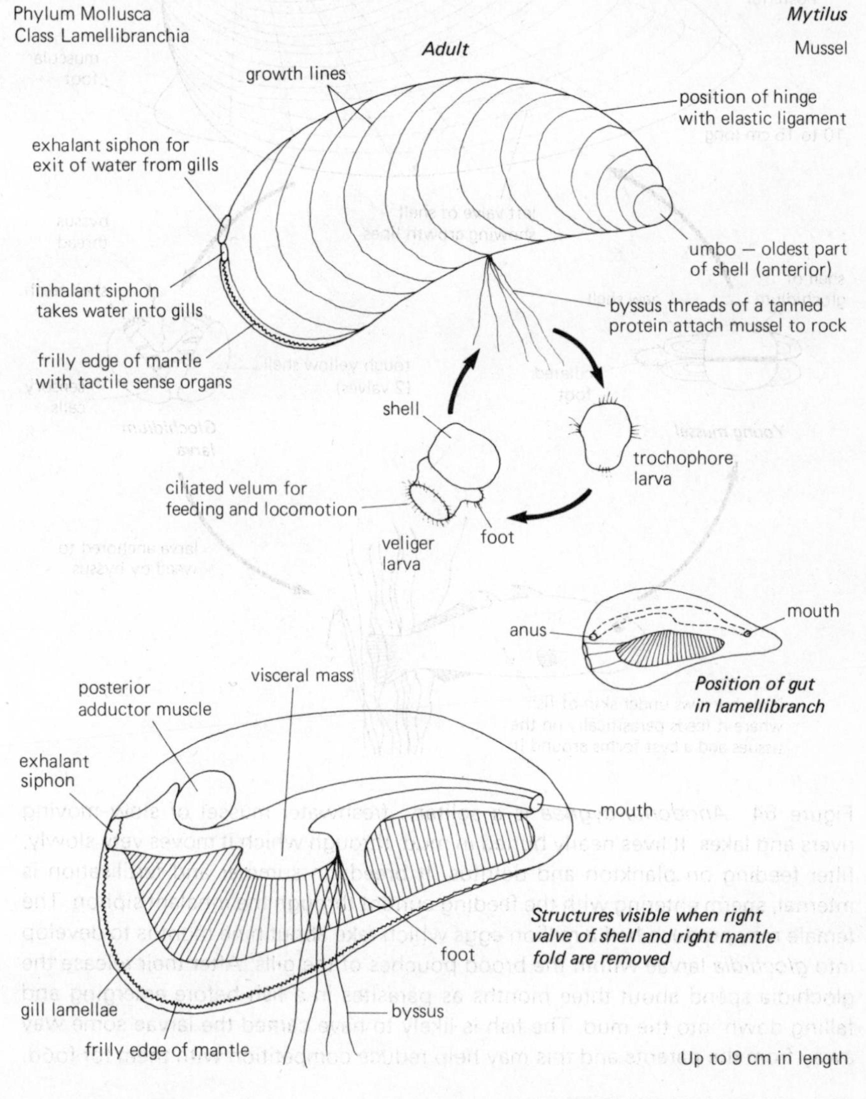

majority of lamellibranchs are sedentary but those that move generally use the foot. Exceptions are the scallops which swim by flapping the two halves of the shell open and shut.

The body is completely enclosed by the two halves of the mantle which secretes the shell. Inside the mantle cavity are *large plate-like gills*, composed of small filaments bearing cilia. The cilia create a powerful current of water which

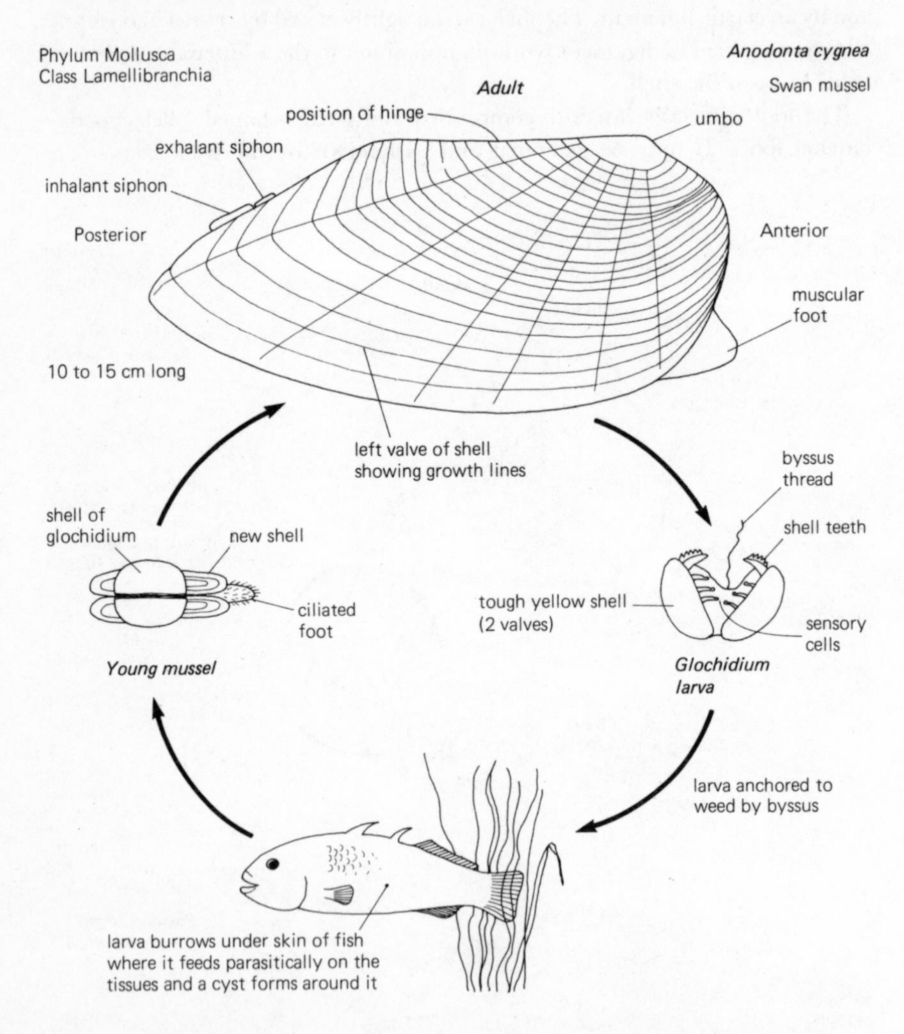

Figure 64 *Anodonta cygnea* is a solitary, freshwater mussel of slow-moving rivers and lakes. It lives nearly buried in mud, through which it moves very slowly, filter feeding on plankton and detritus. It breeds in summer and fertilisation is internal, sperm entering with the feeding current through the inhalant siphon. The female may produce half a million eggs which take about nine months to develop into *glochidia* larvae within the brood pouches of the gills. After their release the glochidia spend about three months as parasites in a fish before emerging and falling down into the mud. The fish is likely to have carried the larvae some way away from the parents and this may help reduce competition with them for food.

brings in food particles. These are strained out by cilia and carried in strings of mucus to the mouth. The water enters through an *inhalant siphon* below the gills and leaves through an *exhalant siphon* above them; both are at the hind end. In bivalves which burrow or bore into wood or rock, the siphons may be elongated by extensions of the mantle cavity. *Mytilus*, a marine mussel and *Anodonta* a freshwater mussel have been chosen as examples (Figures 63 and 64).

Class Cephalopoda

The cephalopods are active, predacious carnivores, having well-developed heads with sense organs, including image-forming eyes. The head is surrounded by four pairs of tentacles in the octopus and five pairs in the squid and cuttlefish, with suckers for seizing the prey. The mouth has strong 'beak-like' jaws as well as a radula.

The shell is internal and very reduced (absent in the octopus). The exhalant

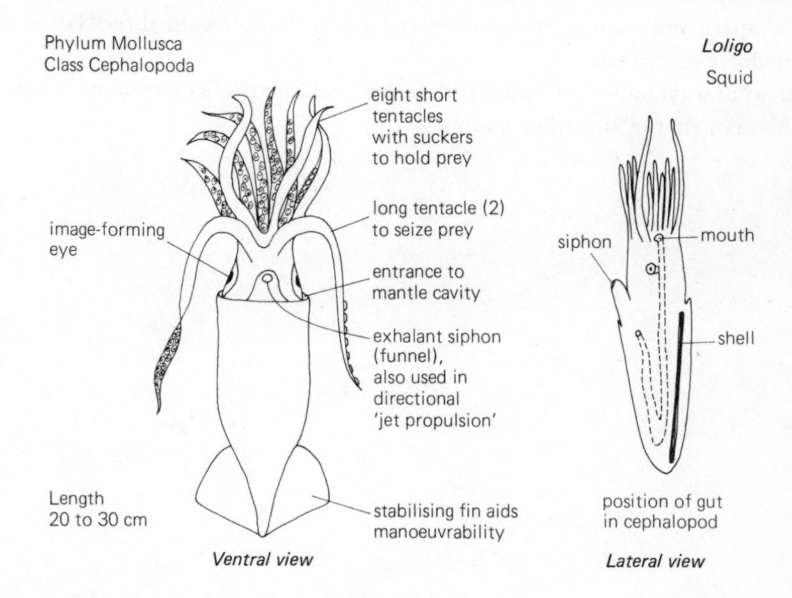

Figure 65 *Loligo forbesi* is a fast-swimming cephalopod of off-shore waters, found near the surface or resting on the bottom in shallow water. It catches shrimps, crabs and fish, using the two long tentacles, then holds the prey with its eight arms and injects poison into it from the mouth. They come together in large numbers at spawning time and there is a complex courtship behaviour. During copulation the male transfers spermatophores to the mantle cavity of the female using its *hectocotylus organ* – a group of modified suckers on the fourth arm on the left side. The fertilised eggs emerge through the funnel of the female and are deposited in communal egg masses inside pods of jelly. The young hatch as miniature adults, but their parents die soon after releasing their gametes. The maximum length of *Loligo* is about 75 cm, but the giant squid *Architeuthis* sometimes reaches more than 16 m to the tips of the tentacles and may account for some 'sea monster' sightings.

siphon or funnel, provides a means of rapid movement by 'jet propulsion'. Most cephalopods can change colour rapidly and can discharge *sepium* – a black substance formed in an *ink sac* near the anus – as a distraction to predators. Squid are streamlined active swimmers, while the broader flatter cuttlefish often rest on the sea floor and octopus with their rounded bodies are inhabitants of rocky coastal waters.

Cephalopods have a complex brain and nervous system and experiments have shown that they have some capacity for learning. *Loligo*, the common squid is described as an example of the group (Figure 65).

QUESTIONS

1. Explain the meaning of the terms mantle cavity, radula, columella.
2. What is torsion? What advantages does it appear to have given to the Gastropoda?
3. Compare and contrast the methods of feeding shown by the three classes of molluscs described.
4. In what ways have the cephalopods become adapted for a more active way of life than that of the other molluscs?

12 Phylum Echinodermata

The echinoderms are marine animals commonly found from the littoral zone to the ocean deeps. There are about 5000 living species and an abundant fossil record. The adult animals are radially symmetrical, usually pentamerous, but the free-swimming larvae are bilaterally symmetrical. They are particularly interesting since they appear to be more closely related to the chordates than are the rest of the invertebrates.

GENERAL FEATURES

They possess a well-developed coelom and are triploblastic. There is a skeleton of calcareous plates embedded just under the surface of the skin, often joined at the edges to form a shell. There may be a covering of spines attached to the plates giving the echinoderms their name – the spiny-skinned animals.

Sea water circulates through a series of canals which make up the *water vascular system* and is used in respiration and locomotion. This system is peculiar to the echinoderms. Attached to the canals are tiny branches called *tube feet*, which can be extended or relaxed by pumping sea water into them or withdrawing it. The tube feet, which often end in suckers, are used in series for walking. A starfish is said to be able to travel up to 15 cm a minute in this fashion.

The nervous system reflects the radial symmetry of the echinoderm body and typically consists of a nerve ring around the mouth and usually five radial nerve cords. The sexes are usually separate and reproduction results in a free-swimming larva.

CLASSIFICATION

There are five classes of present day echinoderms:

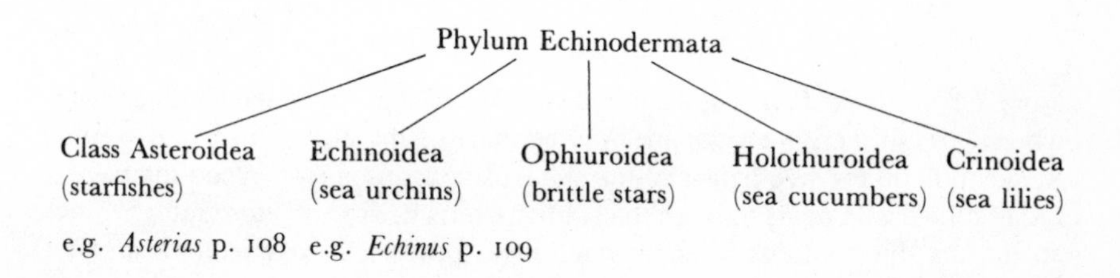

Class Asteroidea

Starfish are typically flattened and star-shaped, with a central disc and a number of arms, usually five. The cushion stars have very short arms, hardly separate from the disc. The mouth of a starfish is on the underside which is therefore called the oral side. *Ambulacral grooves* containing rows of tube feet, run down the underside of each arm and meet at the mouth. Rows of movable spines on each side of the grooves can be bent over to protect the tube feet.

On the upper, aboral, surface the spines are smaller but more numerous. Among them are small tufts of gills through which coelomic fluid circulates. At the bases of the spines are tiny pincer-like structures called *pedicellariae*, for keeping the gills and tube feet clear of debris and also other animals, particularly larval stages about to settle and become sedentary adults. The aboral surface also bears the anus and the *madreporite*, a sieve-like opening to the water vascular system. *Asterias* is illustrated in Figure 66.

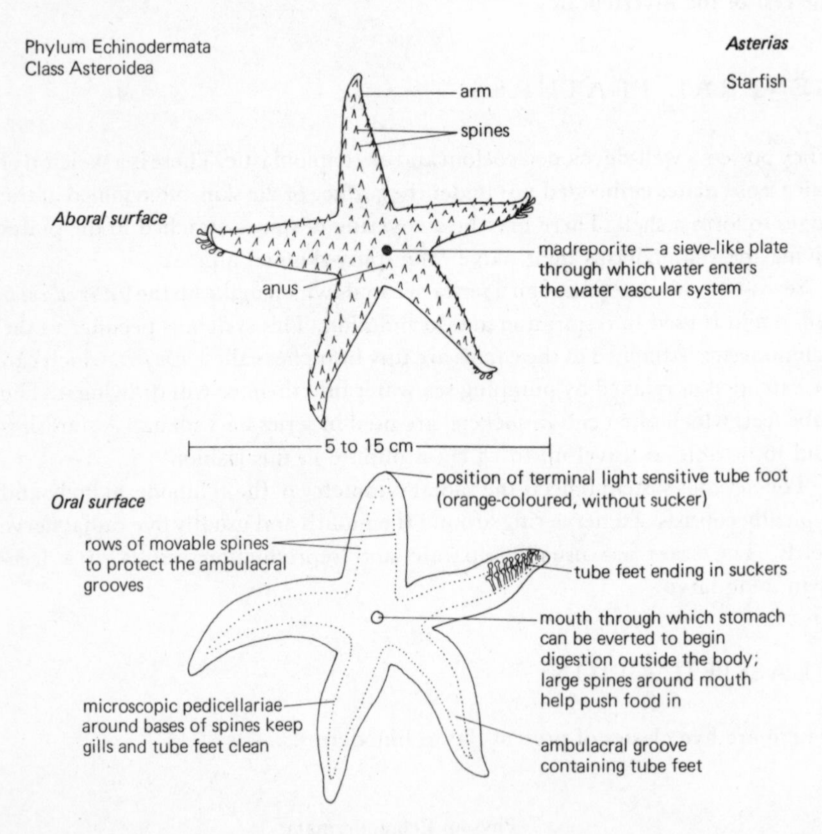

Figure 66 *Asterias rubens* is a common starfish of shallow water, feeding mainly on bivalve molluscs, which it opens by wrapping its arms around them and exerting a steady pull on the two halves of the shell, using its tube feet. When the bivalve tires and the shell begins to open the starfish everts its stomach through the crack and digests the contents. Starfish have great powers of regeneration and are frequently found regrowing lost or damaged arms.

Class Echinoidea

Sea urchins tend to be rounded or heart-shaped. Their skeletal plates are joined to form a shell containing the organs of the body. Most of the plates are covered with spines, sometimes very long for defence, or even with poisonous tips. The tube feet pass out through tiny pores in the shell. The mouth on the underside has five large teeth – part of the jaw apparatus known as *Aristotle's lantern* – the rest of which is hidden inside the shell. The anus and madreporite are on the aboral surface. Pedicellariae of sea urchins are long and used for cleaning and possibly defence. In some species they are used to cover the body with pieces of seaweed presumably as camouflage.

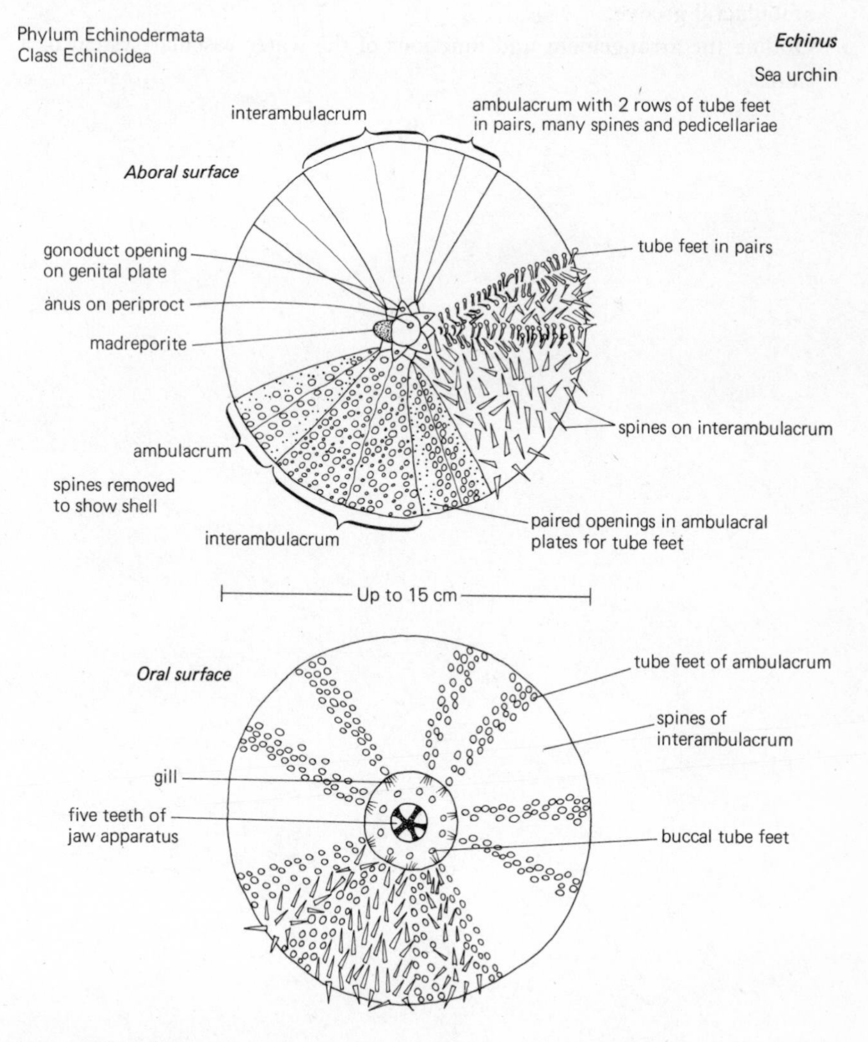

Figure 67 *Echinus* is most common in shallow water on rocks covered in thick growths of *Fucus* and *Laminaria*. It browses on the algae and encrusting animals such as polyzoans. It moves slowly, using its ventral spines and tube feet.

Most sea urchins are found in shallow water, but some, such as the heart urchin *Echinocardium cordatum*, burrow in sand. They live about 20 cm under the surface, with a breathing shaft reaching up to the water above them.

The food of sea urchins may be seaweed in browsing species, such as *Echinus*, or tiny organisms or organic particles collected by the tube feet of *Echinocardium*, or even detritus picked up with sand in some other burrowing forms. *Echinus* is illustrated in Figure 67.

QUESTIONS

1 Explain the meaning of: pedicellariae, madreporite, Aristotle's lantern, ambulacral groove.
2 Outline the arrangement and functions of the water vascular system of a starfish.

13 Phylum Bryophyta

The bryophytes include two small and often neglected groups of plants, the liverworts and mosses. Relatively few have common names and yet they are often very delicate and beautiful, deserving careful examination. They may be found throughout the year in a wide variety of habitats, even in gardens or waste ground in big cities.

In Britain there are about 280 species of liverwort and about 600 species of mosses. They are found in greatest variety where the rainfall and humidity tend to be high, but some, like the wall moss *Tortula muralis*, survive in very dry situations. Other mosses such as *Fontinalis antipyretica*, found in upland streams are entirely aquatic. The liverworts are moisture-loving plants, found at the sides of streams, on damp rocks and amongst wet grasses. A few such as the Floating crystalwort, *Riccia fluitans*, are truly aquatic.

CLASSIFICATION

The bryophytes are divided into two classes. The Hepaticae which are the liverworts and the Musci which are the mosses. The Hepaticae tend to be more flattened in form (thalloid) but some of them, the leafy liverworts do bear small leaves. The mosses tend to be more upright in growth and are leafy.

The classification below deals only with the main orders of each class.

Phylum Bryophyta

Class Hepaticae (liverworts)	Class Musci (mosses)
Order Jungermanniales e.g. *Pellia* p. 112	Order Bryales e.g. *Fanaria* p. 117
Order Marchantiales e.g. *Marchantia* p. 114	Order Sphagnales e.g. *Sphagnum*

Class Hepaticae (liverworts)

The Order Jungermanniales contains the liverworts in which the cells of the thallus are not differentiated for particular functions, such as air chambers. The parts of the thallus where the sex organs develop remain similar to the rest of the plant body without any special structures. This order contains thallose liverworts such as *Pellia* and foliose (leafy) forms such as *Cephalozia*. The leaf-like structures of leafy liverworts are usually divided into lobes which helps to distinguish them from mosses where the leaves are not lobed.

The Order Marchantiales includes the liverworts which show greater differentiation of the thallus, such as *photosynthetic chambers* with pores leading to the exterior. These are visible on the surface of the liverwort as a diamond-shaped pattern with a minute central dot showing the position of the air pore. This group also has its sex organs – *female archegonia* (sing. archegonium) and *male antheridia* (sing. antheridium) carried on stalks above the surface of the thallus. The stalks are termed *archegoniophores* and *antheridiophores* respectively.

Pellia epiphylla

The thallus of *Pellia* is dark green, deeply lobed and shows clear dichotomous branching at its tips. A midrib is usually visible as a darker line running down each lobe. On the underside of the thallus are hair-like *rhizoids* which fix it to the soil (Figure 68).

In early summer small pockets become visible near the tips of a fertile thallus. Inside them are the female reproductive organs, called archegonia. A number of archegonia develop under a flap called the *involucre*. Each archegonium consists of a *neck* leading down to a *venter* at the base in which the female gamete or *oosphere* develops.

The male sex organs, the antheridia, form in groups a little further from the tip. Each contains a mass of *antherozoid mother cells* which divide mitotically to produce haploid *biflagellate antherozoids*. These are released when the antheridium absorbs water and breaks. This ensures that the antherozoids are only released when it is wet enough for them to be able to swim on the surface of the liverwort. They are attracted to the archegonia by a protein secreted by the archegonial necks. When an antherozoid swims down the neck of an archegonium it fuses with the oosphere to form a diploid *oospore*. Normally only one zygote develops under each involucre. The oospore begins its development soon after it is formed and is a well-developed *sporogonium* by the end of the autumn.

It first develops firm contact with the thallus by growing backwards into it. It is then able to absorb food from the thallus and will be dependent on it throughout its existence. The food is absorbed through the lower part of the sporogonium, called the *foot*. Next to this a stalk or *seta* is formed and at the tip a *capsule* develops. Inside the capsule spore mother cells divide meiotically to produce a mass of haploid *spores*. Other cells differentiate into long, thin, spirally thickened structures called *elaters*, grouped together as an *elaterophore*. The whole sporogonium remains inside the archegonium which enlarges to contain it and is then called the *calyptra*.

A resting period now follows until February or March of the next year, then the wall of the capsule becomes thickened and the seta elongates, rupturing the calyptra and carrying the capsule above the thallus. The capsules are round,

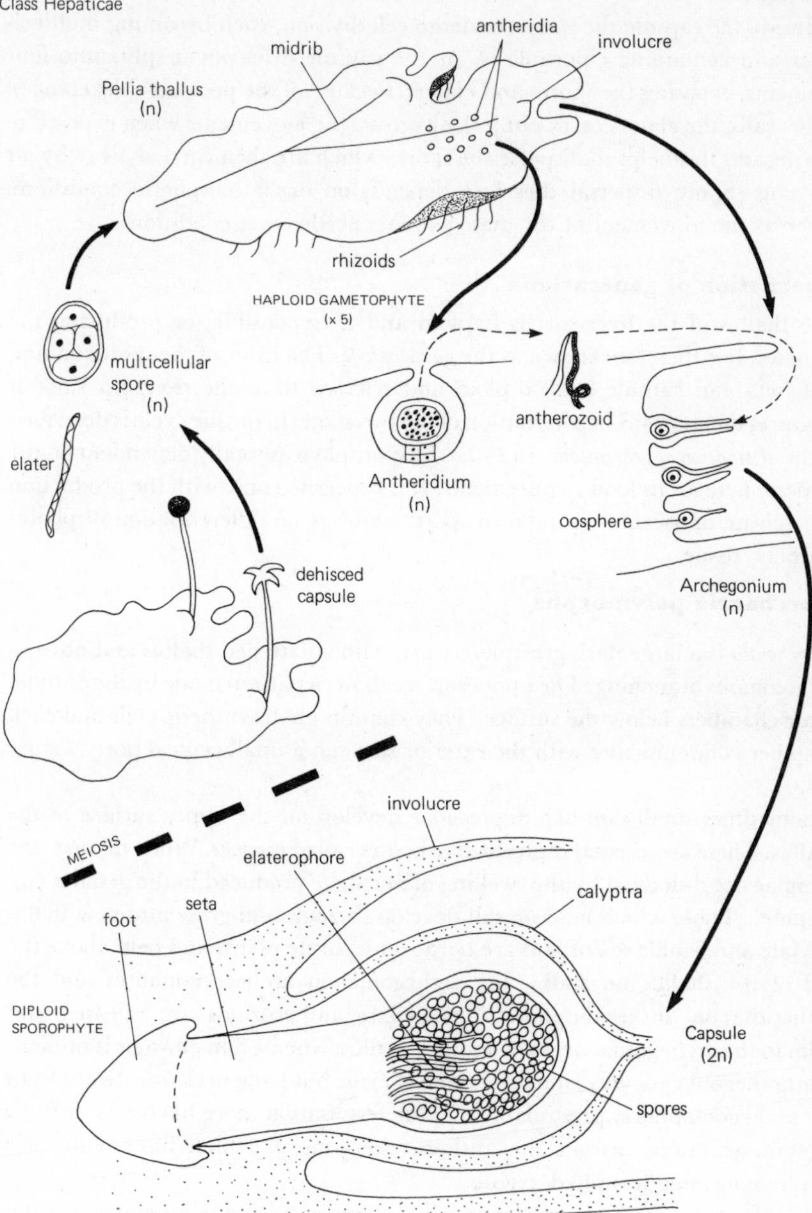

Figure 68 Life cycle of *Pellia*
Pellia epiphylla is a typical thallose liverwort showing very little differentiation of its tissues. It may be found growing in damp, shady situations, often on the banks of streams and may cover a large area with its dark green prostrate thallus. Its very smooth surface and absence of gemmae cups make it distinctive from other similar large liverworts.

black and shiny, showing up clearly on their white stalks, which may reach 5 cm in height.

Inside the capsule the spores undergo cell division, each becoming multicellular and containing chloroplasts. As the capsule dries out it splits into four segments, exposing the spores and elaters. Because of the peculiar thickening of their walls, the elaters carry out jerky hygroscopic movements when exposed to the air and this helps to disperse the spores which are then carried away by air currents. Spore dispersal therefore depends on dry atmospheric conditions, whereas the movement of the male gametes needed wet conditions.

Alternation of generations

The thallus of the liverwort is haploid and is responsible for producing the gametes. It is therefore known as the *gametophyte*. The tissue of the sporogonium, foot, seta and capsule is all diploid and referred to as the *sporophyte* since it produces the spores. The alternation of the two stages in the life cycle is described as the *alternation of generations*. In *Pellia* the sporophyte is totally dependent on the gametophyte for its food requirements. It is concerned only with the production and release of asexually-produced spores and has no differentiation of photosynthetic tissue.

Marchantia polymorpha

Marchantia is a large dark green liverwort, with a flattened thallus and obvious dichotomous branching. The upper surface shows a pattern made by the outlines of air chambers below the surface. They contain photosynthetic cells and each chamber communicates with the exterior through a small central pore (Figure 69).

Sometimes small cup-like depressions develop on the upper surface of the thallus. These are asexual reproductive bodies called *gemmae*. When mature, the gemmae are dislodged by the swelling of mucilage produced in the gemma cup (cupule). Those which land on soil develop rhizoids and grow into new thalli.

Male and female sex organs are borne on separate plants and held above the rest of the thallus on stalks, the archegonia on archegoniophores and the antheridia on antheridiophores. Biflagellate antherozoids are released and swim to the archegonial necks of another thallus, when a film of water is present. The archegonia are very similar to those of *Pellia* but hang neck downward from the archegoniophore, presumably making fertilisation more likely. As in *Pellia* protein secretions attract the antherozoids, one of which fuses with each oosphere forming a diploid zygote.

The zygote undergoes a series of cell divisions resulting in an embryo of the typical liverwort plan with a foot embedded in the thallus, a seta and a capsule. The foot absorbs food from the thallus. Inside the capsule haploid spores are produced by meiosis. Spindle-shaped elaters are also present which coil and uncoil with variations in humidity when the capsule splits open. If the spores land in a suitable place they will germinate to produce a new haploid thallus. Once again there is clear alternation of a haploid gametophyte and a diploid sporophyte, which is dependent on the gametophyte for its food.

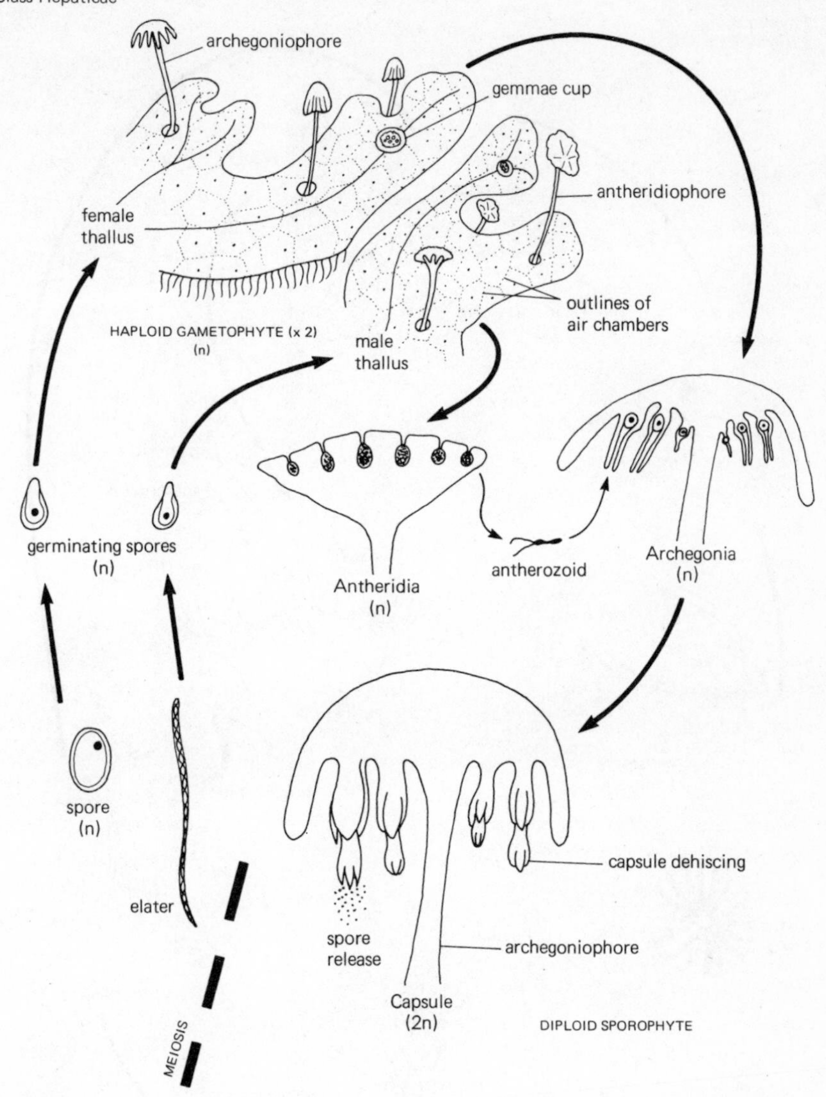

Figure 69 Life cycle of *Marchantia*
Marchantia polymorpha is one of the largest British liverworts. A single thallus may reach more than 7 cm in length. It is found in moist positions on banks and rocks. The pattern of air chambers and rounded gemmae cups makes it easily recognisable.

Class Musci (Mosses)

Moss plants tend to be found in groups rather than as individuals, often forming tussocks or little cushions. A single plant may be only a few centimetres tall or up to 25 cm in the case of the tallest hair mosses, *Polytrichum*. The stems may be

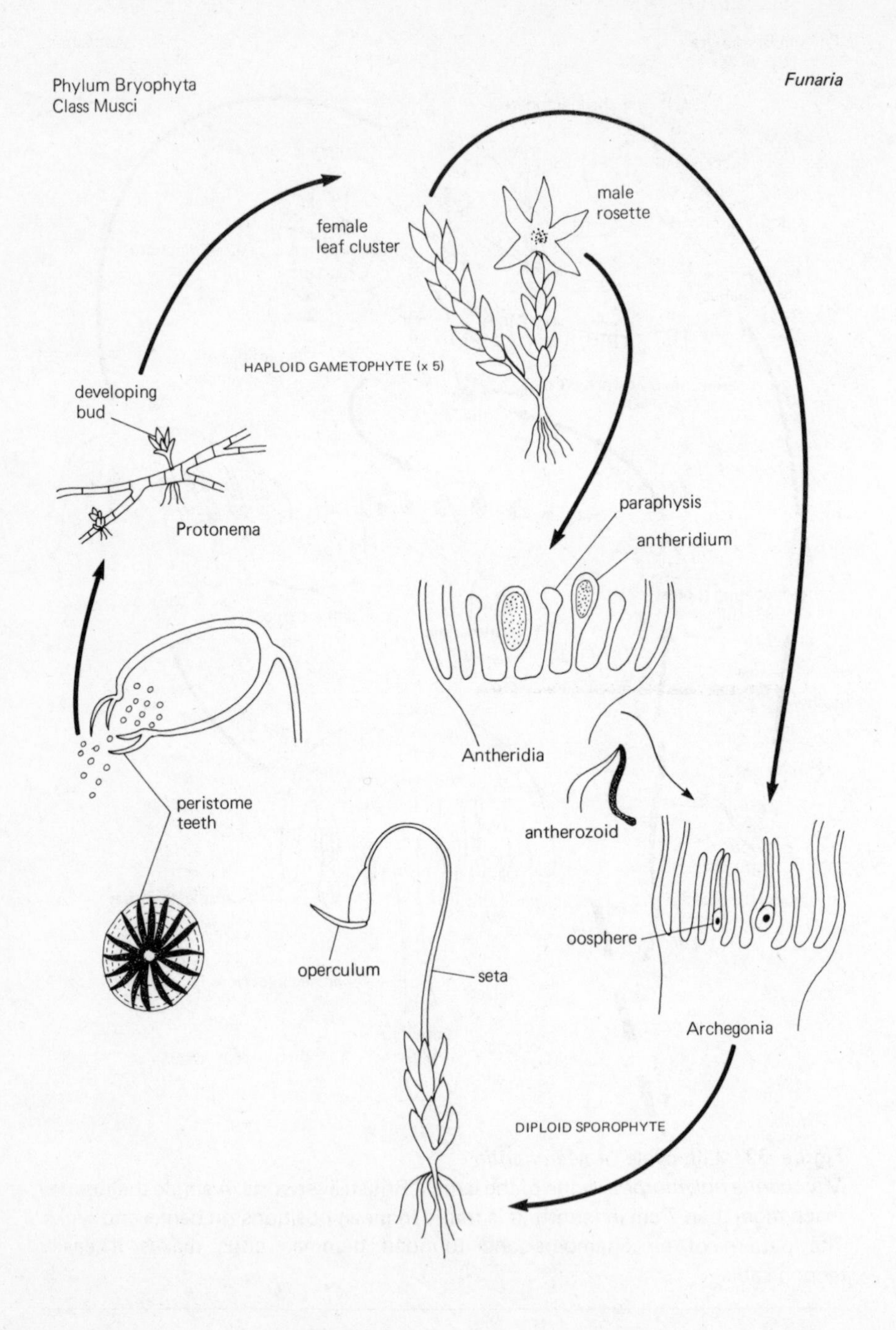

Figure 70 Life cycle of *Funaria*
Funaria hygrometrica is a common moss found growing in patches on damp soil or on walls. It is often found on soil which has been burnt after a forest fire, or even on paths made from cinders. Each plant grows about 2 cm tall and is covered in spirally-arranged yellowish-green leaves.

upright or prostrate and usually have many branches, forming tufts, as in *Tortula*, or hanging downwards as in the bog mosses, *Sphagnum*. *Sphagnum* is a commercially important moss used in horticulture to help soil to retain its moisture.

The mosses show greater differentiation of tissues than in the liverworts, including the beginnings of simple conducting strands for food and water. There is a definite stem, usually with three or more rows of spirally arranged leaves. The leaves are never lobed but often have a well-developed midrib.

Although the terms, leaves, stems etc. are used for convenience, these bryophyte structures are not homologous with the stems and leaves of the higher plants. Bryophyte stems and leaves are borne on a haploid gametophyte, while those of the higher plant are part of a diploid sporophyte.

Funaria hygrometrica

The *Funaria* plant has oval leaves with a midrib running their whole length. At the base of the stem are numerous slender brown multicellular rhizoids which anchor the plant in the soil. There are comparatively few branches and they grow off from below the leaves, not from their axils (Figure 70).

Funaria is *monoecious*, having both antheridia and archegonia on the same plant. The antheridia are formed at the end of the main axis, while the archegonia are developed at the end of a lateral branch. The leaves around the antheridia become spread out like a rosette and turn brown. They are called *perichaetia* and are much more conspicuous than those around the archegonia.

The antheridia are essentially the same as those of the liverworts but the archegonia have rather longer necks and are interspersed with sterile hairs called *paraphyses*. The biflagellate antherozoids are attracted to the necks of the archegonia by secretions of sucrose. Fertilisation results in an oospore which develops into a sporogonium with foot, seta and capsule. The seta elongates to about 3 cm holding up the capsule for spore release. The capsule is much more complex than that of a liverwort. It contains some sterile tissue not involved in spore production but concerned with other activities such as photosynthesis. However the sporophyte remains dependent upon the gametophyte for nearly all its food supply.

Spore release does not involve elaters, but the action of a complex system of *peristome teeth*. The top of the capsule is protected for a time by an *operculum*, which is eventually forced off when the *annulus* cells below it absorb water. The peristome teeth are then exposed to the atmosphere. In *Funaria* there are sixteen inner and sixteen outer lignified teeth. Variations in humidity cause them to make hygroscopic movements, bending outwards when it is dry and inwards when wet. As in the liverworts, spore dispersal is thus restricted to dry atmospheric conditions.

Moss spores are very tiny and easily carried by wind. Under suitable conditions they germinate to produce a branched filamentous structure called a *primary protonema*. Cell divisions of the protonema form buds which develop eventually into new moss plants.

Funaria does not produce gemmae as do many mosses, but it does show asexual reproduction by the formation of filamentous growths from any part of the gametophyte. These are distinguished as *secondary protonemata*, separate from the primary protonemata derived from spores.

QUESTIONS

1 What are the main morphological differences between a liverwort and a moss?

2 Draw a diagram to show the main stages in a bryophyte life cycle. Use it to help explain the terms gametophyte, sporophyte and alternation of generations.

3 In what ways do you consider a moss better adapted than a liverwort for terrestrial conditions?

14 Phylum Pteridophyta

The pteridophytes include the ferns, club mosses and horse-tails. They are a very ancient group of plants, forming the dominant vegetation of the Paleozoic era, from the Upper Devonian, 260 million years ago to the end of the Lower Permian about 200 million years ago. Later, first the gymnosperms and then the flowering plants evolved and largely replaced them.

They are often described as the *vascular cryptogams*. The term vascular refers to the fact that the pteridophytes have well-developed vascular tissue, xylem for water conduction and phloem for food conduction. This is a considerable advance compared to the very simple elongated conducting cells sometimes found in the bryophytes. The term cryptogam refers to the reproductive processes of the pteridophytes which, before microscopes were developed, were considered to be hidden as compared to the obvious flowers of the flowering plants.

Like the bryophytes, the pteridophytes show a very clear alternation of generations, but the dominant generation is the diploid sporophyte, which develops root and shoot systems homologous with those of the flowering plants. It soon becomes independent of the gametophyte generation which remains a small structure, a few centimetres or less across and usually dies soon after the development of the young sporophyte. The sporophyte is well-adapted to life on land, but the gametophyte is often a thin plate-like structure, susceptible to dehydration. Some gametophytes however are subterranean – see *Lycopodium* p. 123 or retained within a resistant spore wall – see *Selaginella* p. 125 and able to survive in a wider range of habitats.

Where the gamete remains inside a spore the spores are of two different sizes (*heterosporous*). The large *megaspores* give rise to female prothalli bearing archegonia and the smaller *microspores* give rise to male prothalli bearing antheridia. *Selaginella* is an example of a heterosporous pteridophyte, whereas the ferns and some club mosses, such as *Lycopodium*, are *homosporous*, producing only one type of spore.

Retention of the gametophyte within the spore gives it some protection but allows little or no photosynthesis and makes food storage necessary, especially for the megaspores in which the embryos will develop.

Separation of the antheridia and archegonia on different prothalli appears to give both advantages and disadvantages. It obviously prevents self-fertilisation and the greater variation allowed by cross-fertilisation may well have had advantages for survival in a changing environment. On the other hand it makes fertilisation seem less likely since the gametes may be produced at some distance from each other. The small size of the microspores (which do not have to have food stores) compensates for this since they can be blown a long way by wind, and very large numbers can be produced from a small amount of the sporophyte tissue. Both factors increase the likelihood that they may fall on to a female prothallus and release antherozoids which swim, in a thin film of water, to the neck of an archegonium. The wind-blown microspore is also unlikely to have come from the same sporophyte as the female prothallus on which it happens to fall, so that still more variation is made possible.

Evolution of heterospory is believed to be very important in the development of the seed habit which we shall be considering later in relation to the gymnosperms and the flowering plants. Heterospory and homospory are also important in the classification of the pteridophytes because reproductive organs are usually a good guide to the relationships between plants. On the whole they are less subject to changes brought about by the environment than are the vegetative parts so that resemblances between reproductive organs of different plants suggest that they are likely to be related phylogenetically.

CLASSIFICATION

Fossil forms of Pteridophyta are very common, especially in rocks of the Carboniferous period and any attempt to classify modern pteridophytes has to take into account their fossil record. Unfortunately the fossils generally consist only of fragments of plants, making it difficult to reconstruct the whole plant or to work out its relationship with another. It also becomes necessary to revise the classification from time to time as more fossil evidence becomes available. Some sub-classes contain only fossil representatives, others a mixture of both fossil and living forms. The classification given here is a simplified one, including only those groups which contain living examples of British Pteridophyta and omitting those groups which are represented only by fossils or mainly by living tropical species.

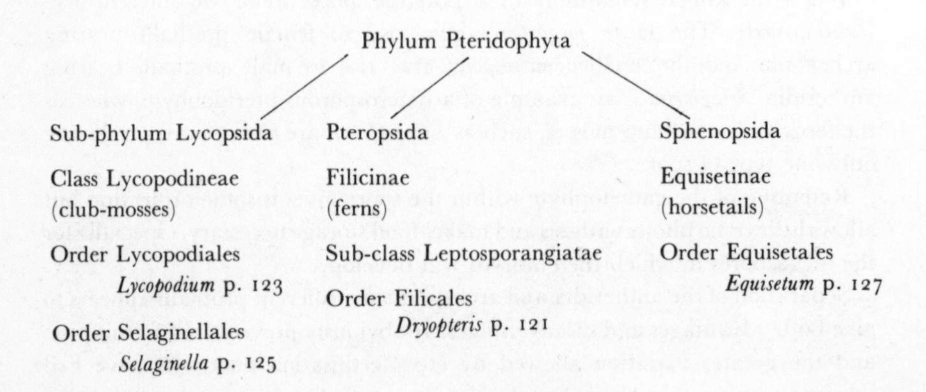

Class Filicinae

Dryopteris filix-mas (the Male fern)

The Male fern plant is the sporophyte generation and therefore has true stems, roots and leaves (Figure 71). It has an underground stem axis, bearing tough fibrous adventitious roots which attach it very firmly to the soil. The older parts

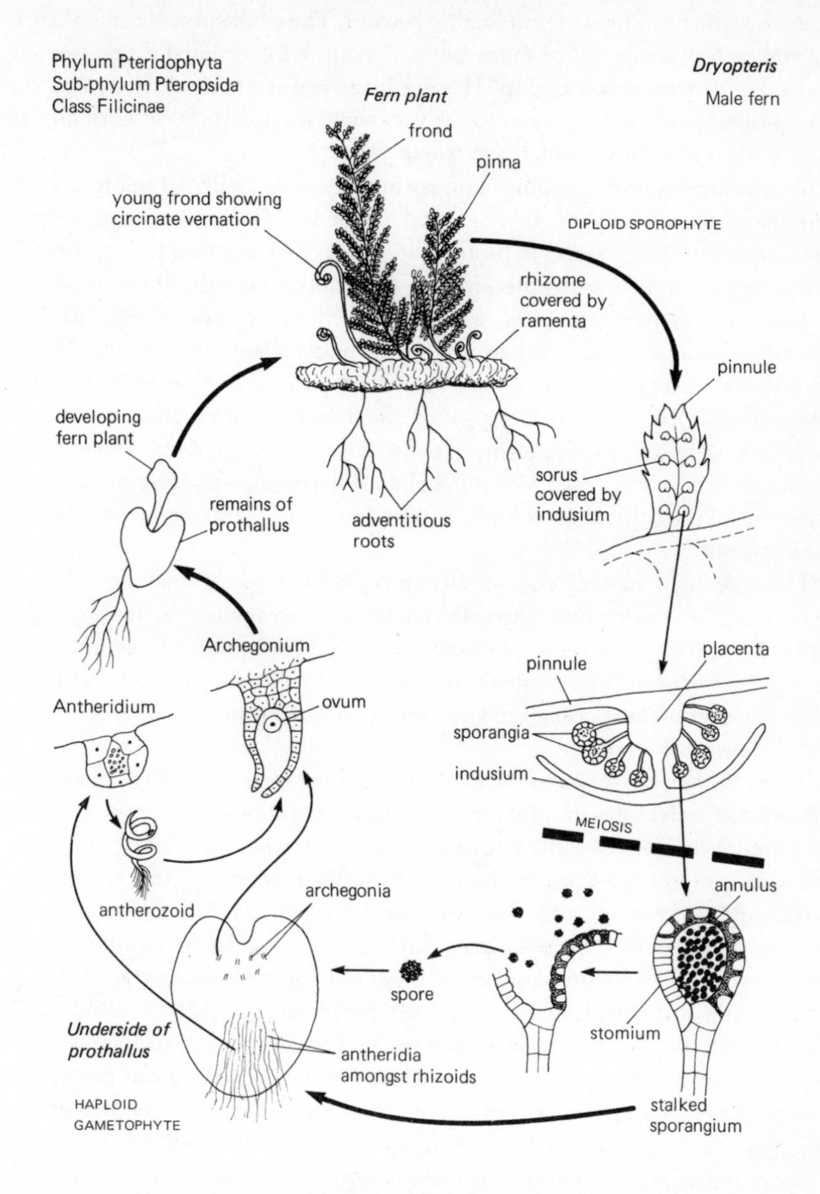

Figure 71 Life cycle of *Dryopteris*
Dryopteris filix-mas – the male fern – is common all over the British isles in woods, hedgerows and open heathland. Its leaves are lance-shaped, tapering to a sharp point and reaching a metre or more in height.

of this rhizome gradually wither and die as younger growths develop. If side branches are produced they may eventually become separated from the main axis and form separate plants. Above ground there are large leaves or fronds, their bases surrounded by dry scales called *ramenta*. This covering protects the delicate young leaves from frost or drought. The fronds are composed of a main axis called the *rachis* and numerous leaflets on either side called the *pinnae*. Each pinna is further subdivided into smaller *pinnules*. The whole frond is described as a *compound leaf* because of its many parts. A mature frond may be a metre long and takes two years to develop. Having large leaves or fronds is termed the *macrophyllous condition* as opposed to the very small leaves of other Pteridophytes, such as the club mosses which are *microphyllous*.

Very young leaves developing in the spring are coiled up like a bishop's crozier and uncoil slowly as they grow. This kind of leaf folding is called *circinate vernation* and probably serves to protect the young emerging leaf. It is a primitive characteristic found in many pteridophytes and absent from the Spermatophyta.

The reproductive structures which produce the spores of the fern are sporangia borne in clusters called *sori*, on the undersides of the pinnules. These are ordinary vegetative leaves but may be called *sporophylls* because they bear the spores. Each sorus has a covering called the *indusium*, under which a group of sporangia develop. Inside them, *spore mother cell* nuclei divide meiotically producing forty-eight or sixty-four unicellular, haploid spores. These will be the beginning of the haploid gametophyte generation. All the spores are identical – homosporous.

The sporangia have a curious structure, related to the method of spore dispersal. They are round, laterally compressed structures on multicellular stalks. Around the edge of the sporangium is a row of cells called the *annulus*, each of which is thickened on its radial and inner walls. The annulus stops half way down one edge of the sporangium and here the cells are thin-walled, making up a region called the *stomium*.

When the spores are mature in late summer the indusium shrivels and finally falls off. The cells of the annulus lose water by evaporation now being exposed to the atmosphere. Loss of water causes their cell contents to shrink away from the cell walls and exerts an inward pull on the walls themselves. Only the thinner outer walls can bend inwards, however, and as they do so strains are set up in the annulus. Finally, the cells of the stomium rupture, allowing the annulus to curl back on itself. This sudden movement flings out many of the spores. As drying continues the shrinking contents of each cell part from its wall and a bubble of air appears in each cell. The tension is then released and the annulus suddenly snaps back from its wound up state to its original position, throwing out any spores which are left. Spores may be shot out several centimetres by this method and in one season a large plant of *Dryopteris* may produce several million spores.

Spore release has obviously depended on warm, dry conditions, but the next stage in the fern's life cycle – the germination of the spore, needs a moist situation.

The spores remain dormant for a while then germinate, putting out a green cell and then a colourless rhizoid, for anchorage and absorption. The green cell divides many times and eventually forms a heart-shaped, nearly flat plate of cells called a *prothallus*. This is the gametophyte generation and all its cells are haploid.

The gametophyte is about one centimetre across and anchored to the soil by many unicellular rhizoids, developed under the central region of the plate where it is several cells thick. The cells contain chlorophyll and carry out photosynthesis, but there is no division of labour, no conducting tissue and no protecting layer of cells at the surface. Thus the prothallus is very delicate and vulnerable to desiccation, only surviving in damp surroundings.

Male and female reproductive organs are developed on the prothallus, the male antheridia form on the underside amongst the rhizoids and the female archegonia develop just under the notch of the prothallus. The archegonia are very similar to those of the bryophytes. Each consists of a neck leading down to the oosphere, the female gamete, inside the venter at the base of the archegonium.

Inside each antheridium thirty-two antherozoid mother cells develop, without further division, into flagellate antherozoids which are then released and swim to the archegonia of another prothallus. They are attracted by malic acid (2-hydroxybutane dioic acid) secreted by the archegonium. Since the antherozoids normally mature before the archegonia, self-fertilisation is very rare, but their passage from one prothallus to another requires a film of water in which they can swim.

Fertilisation of the oosphere results in an oospore which is the first stage of the sporophyte generation. It begins development inside the remains of the archegonium, forming first a root, leaf, stem apex and an absorptive foot. The foot becomes embedded in the tissue of the prothallus, from which it absorbs nourishment. The first root anchors the young sporophyte but is soon replaced by adventitious roots developing from the stem axis. The first bilobed leaf appears bending up through the notch of the prothallus. As the young sporophyte gradually becomes an independent fern plant the prothallus withers away.

Class Lycopodineae

Lycopodium (a club moss)

The *Lycopodium* sporophyte has branching stems which may be erect (*L. selago*) or trailing (*L. clavatum*) and are completely covered in small spirally-arranged leaves. The undersides of the stems bear adventitious roots at intervals (Figure 72).

Lycopodium, like *Dryopteris*, is homosporous and the sporangia are borne on the upper surfaces of certain leaves or sporophylls. In *L. selago* there are alternating regions of fertile and sterile leaves along the stem; but in *L. clavatum* the sporophylls are slightly different in shape from the other leaves and are grouped together as *cones* or *strobili* on special aerial branches called *podia*. *L. selago* also reproduces vegetatively by means of detachable bulbils produced just behind the apices of young shoots. Each *bulbil* is a small leafy lateral stem which occurs in place of a leaf.

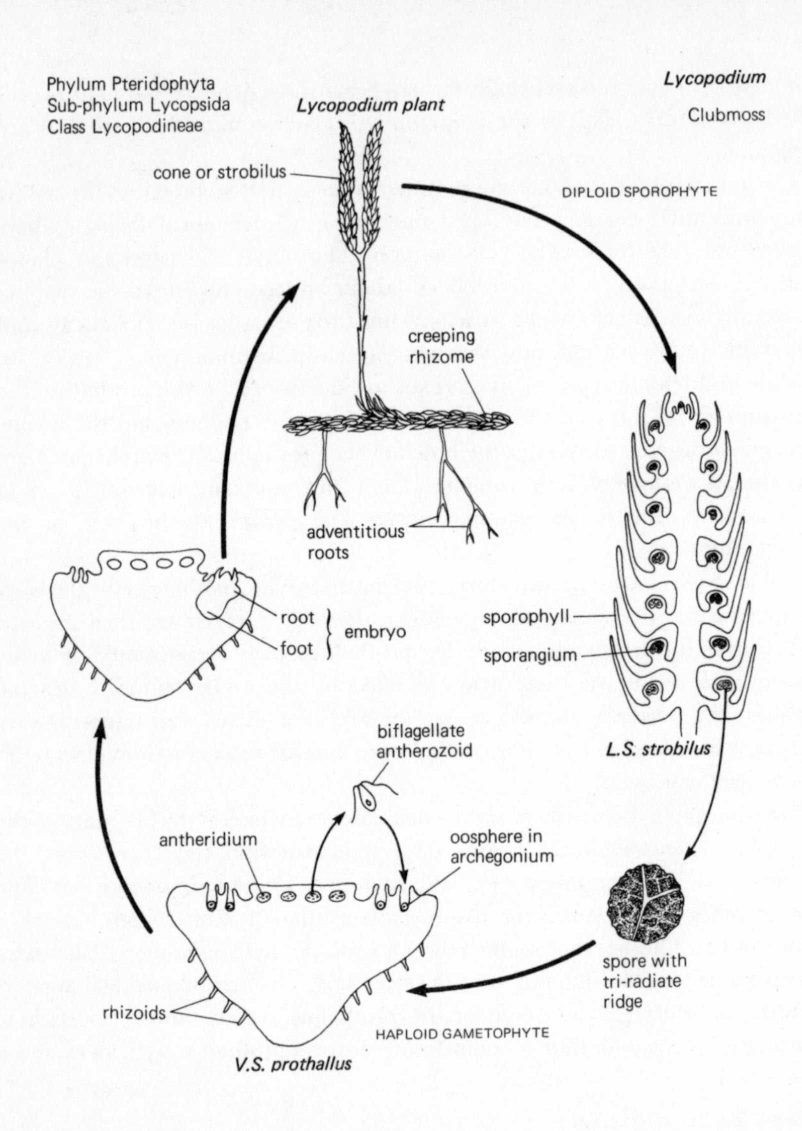

Figure 72 Life cycle of *Lycopodium*
Lycopodium clavatum – this club moss is one of five British species, found on high moorland in Wales and Scotland. Its branching stems have a creeping habit and support aerial cone-bearing branches which may reach about 12 to 15 cm in height.

The spores of most British species of *Lycopodium* remain dormant for several years, sometimes as long as six or seven, before germinating. The prothalli which then develop are one to two centimetres in diameter and are largely subterranean. The upper layers of *L. selago* prothalli may be at soil level and a little chlorophyll may develop; otherwise it, like other *Lycopodium* prothalli, receives nourishment from a mycorrhizal fungus which must enter it at an early stage or it will die. The fungus and the prothallus appear to have a symbiotic relationship, that is, one which is of mutual benefit to them, but its exact nature is not clear.

The gametophyte may take several years to mature and may take as long as fifteen years to do so, but eventually it produces antheridia and archegonia. A biflagellate antherozoid fertilises an oosphere in one of the archegonia and the resulting embryo is the beginning of a new sporophyte plant. An absorptive foot is produced as it was in *Dryopteris* and for a long time the developing sporophyte obtains nourishment from the gametophyte. At last, when it is sufficiently well developed it becomes fully independent.

Selaginella

The *Selaginella* sporophyte has four rows of leaves on its creeping stem. In *S. kraussiana* the two rows of leaves on the upper surface are small, while the two rows on the lower surface are larger. In *S. selaginoides* all the leaves are the same size. On the upper surface at the base of each leaf there is a small membranous structure called a ligule. The fast growth of the side branches makes them appear falsely dichotomous. (See Figure 73.)

The stem of *S. selaginoides* bears roots at intervals, but *S. kraussiana* has curious structures called *rhizophores* which grow downwards and develop tufts of roots at their tips. These structures are peculiar to this order and have no parallel in the rest of the plant kingdom since they differ in structure and habit from both stems and roots.

The sporangia are produced on sporophylls near the apices of fertile erect shoots. The sporophylls are very similar to foliage leaves. Two kinds of sporangia are distinguishable – *megasporangia* which produce four large *megaspores*; and *microsporangia* which produce a large number of small *microspores*. *Selaginella* is therefore heterosporous. The same cone or strobilus usually bears both kinds of sporangia, with the microsporangia generally nearer the top.

The two different kinds of spore develop into two different kinds of gametophyte. The microspore begins its development before it is shed from the sporophyte and its contents become a male gametophyte consisting of a single antheridium. Inside it 128 biflagellate antherozoids develop.

Inside each megaspore a female gametophyte also begins its development before it is released. The spore wall has three ridges and eventually ruptures where these ridges meet. The split extends along the ridges exposing the top of the prothallus, where archegonia begin to form. The top part of the prothallus may also bear rhizoids and develop chlorophyll, but the bulk of its nourishment, like that of the male prothallus, came from the sporophyte. In the megaspore this food is contained in a special food storage region in the lower part of the spore. The gametophytes of *Selaginella* are not nutritionally independent of the sporophyte as they are in *Dryopteris* and *Lycopodium*.

The walls of the microspores also rupture along their three ridges, releasing the antherozoids. They are biflagellate and swim to the archegonia of a female prothallus where fertilisation of an oosphere occurs. The separation of the male and female prothalli makes self-fertilisation impossible.

The fertilised oospore marks the beginning of the new sporophyte generation.

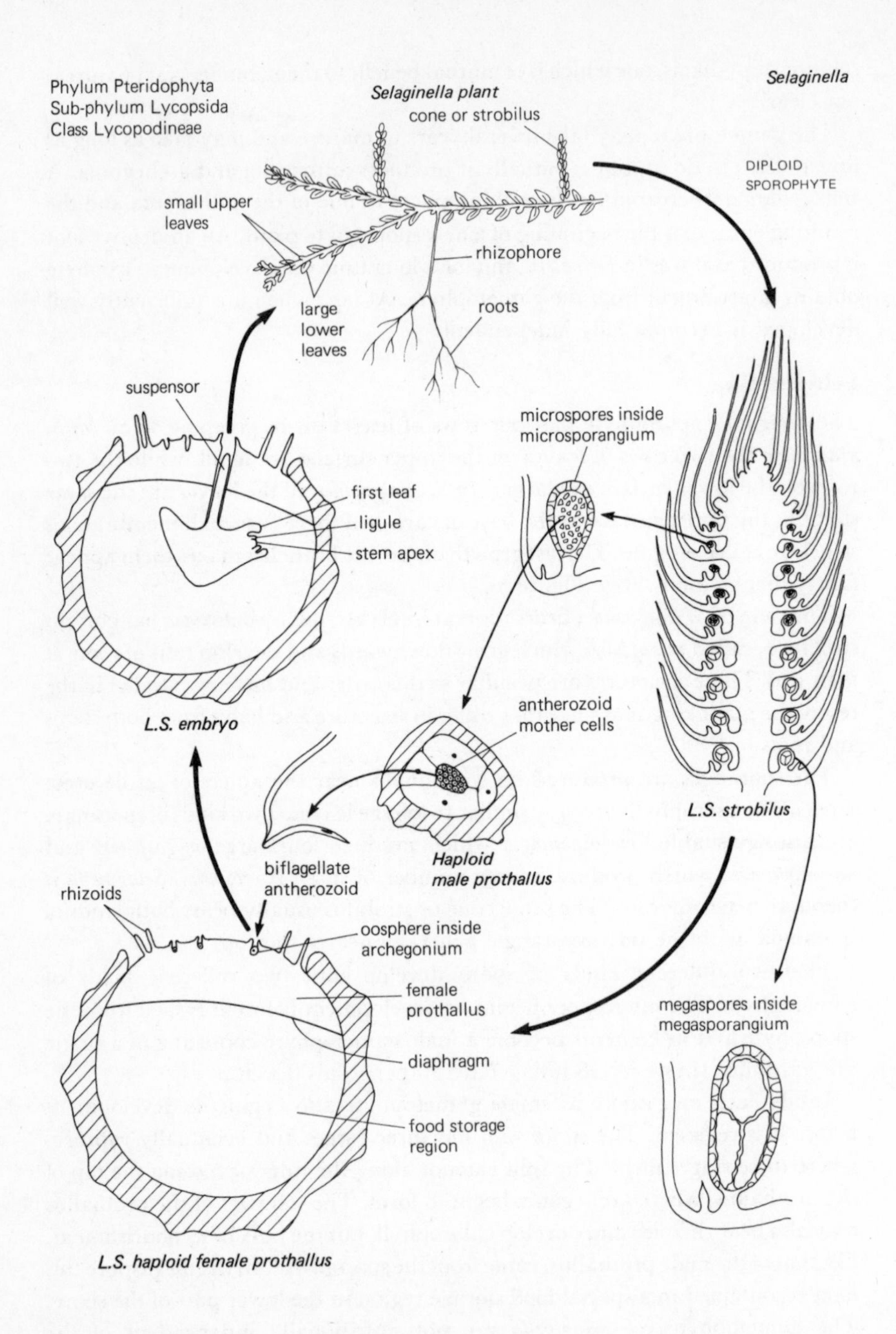

Figure 73 Life cycle of *Selaginella*
Selaginella kraussiana – this species is a very common greenhouse plant and has now become naturalised locally in this country. Its creeping, much branched stems support erect cone-bearing shoots a few centimetres high. The only British species is *S. selaginoides* found on wet highland pastures in Scotland and Wales.

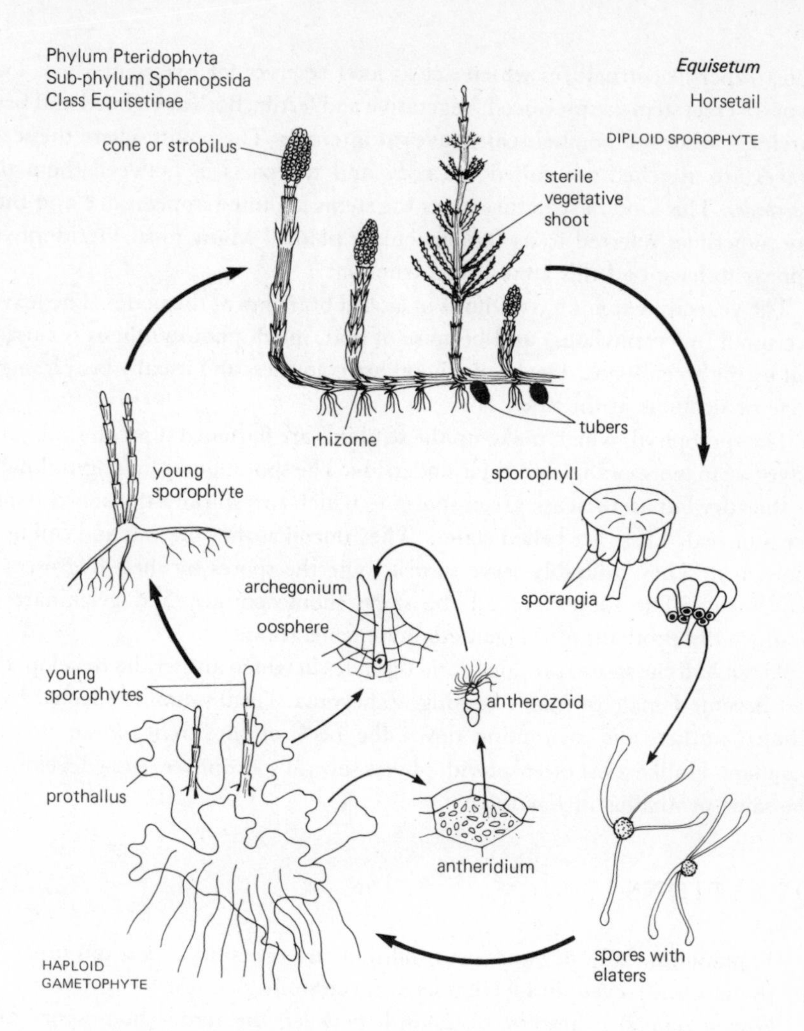

Figure 74 Life cycle of *Equisetum*
Equisetum arvense – the Common horsetail or Field horsetail is found in drier conditions than most other horsetails, many of which are marsh or water plants. It is one of the larger species, but rarely reaches a metre in height.

The upper part of the embryo becomes a *suspensor* which pushes the embryo down through the diaphragm into the food storage region of the megaspore. Eventually the sporophyte becomes fully independent and the gametophyte, now without its food store, dies and disintegrates.

Class Equisetinae

Equisetum (a horsetail)

The Common horsetail, *Equisetum arvense*, has a branching underground rhizome on which erect stems are produced annually (Figure 74). On the rhizome are short tuber-like structures which act as food reserves for overwintering. Two types of erect stem are produced, vegetative and fertile. Both are ribbed and bear circles or whorls of pointed scale leaves at intervals. The points where the scale leaves are attached are called the *nodes* and the parts in between them the *internodes*. This kind of structure gives the stems a jointed appearance and they are sometimes referred to as the 'articulate plants'! Many fossil Pteridophyta appear to have had this kind of construction.

The vegetative stems have whorls of lateral branches at the nodes. The leaves are small (microphyllous) and because of that, much photosynthesis is carried out by the green stems. The fertile shoots are colourless and usually bear a single cone or strobilus at the apex.

The sporophylls which make up the strobilus are flattened discs on stalks and have five to ten sporangia on their underside. The sporangia split longitudinally as they dry out and release green spores to which two spirally-thickened bands are attached. These are called elaters. They uncoil as they dry out and coil up if moistened. They probably serve to distribute the spores by their hygroscopic movements. The spores are all the same (homosporous) and germinate to produce the prothalli of the gametophyte generation.

About half the spores produce gametophytes in which antheridia develop; the rest become female prothalli bearing archegonia. Fertilisation is effected by a ciliated antherozoid swimming down the neck of the archegonium to the oosphere. Unlike most other pteridophytes several sporophytes may develop on the same prothallus in *Equisetum*.

QUESTIONS

1. Explain how both dry and wet conditions are necessary at various times for the fern's life cycle to be completed successfully.
2. Giving named examples, distinguish between the terms: homospory and heterospory; microphyllous and macrophyllous.
3. In what ways may the pteridophytes be considered better adapted to terrestrial life than are the bryophytes?
4. Compare and contrast the life history of a named pteridophyte with that of a named bryophyte.

15 Phylum Spermatophyta – Introduction

Most of the plants we see around us are spermatophytes or seed plants, varying in size from trees and shrubs (see p. 138) to small flowering plants like dandelions and grasses. They begin their life cycle as seeds produced either inside cones as in fir trees and pines, or inside flowers.

To see how seeds could have evolved we have to look at both fossil and living representatives of the Pteridophyta and Spermatophyta. Unfortunately there are many gaps in the fossil record because soft plant tissue decays very quickly. Often there are only fragments of leaf or stem to refer to and these may show similarities with both groups. Unless reproductive organs are found with them, it may be impossible to classify the remains accurately. However, we can begin to seé how the changes may have occurred.

There has been a gradual reduction in size, complexity and independence of the gametophyte. In the Bryophyta – the liverworts and mosses – the gametophyte is the plant body and it supports a sporophyte, consisting of a capsule, inside which spores develop (Figure 75a).

The Pteridophyta – the ferns, club mosses and horsetails (see chapter 14), also demonstrate a clear alternation of gametophyte and sporophyte generations, but here the sporophyte is well-adapted to terrestrial conditions, whereas the gametophyte is more delicate. It is susceptible to desiccation and requires a film of water on its surface for fertilisation by swimming male gametes, (Figure 75b).

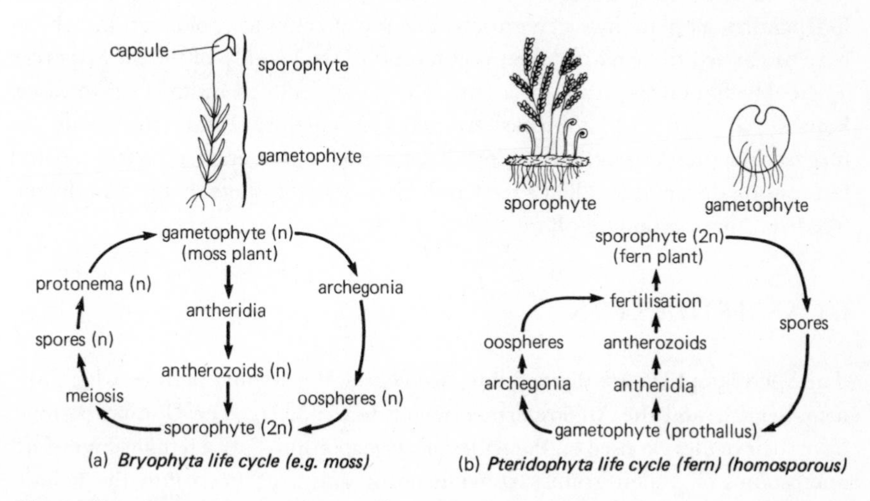

Figure 75 Comparison of plant life cycles, (a) and (b)

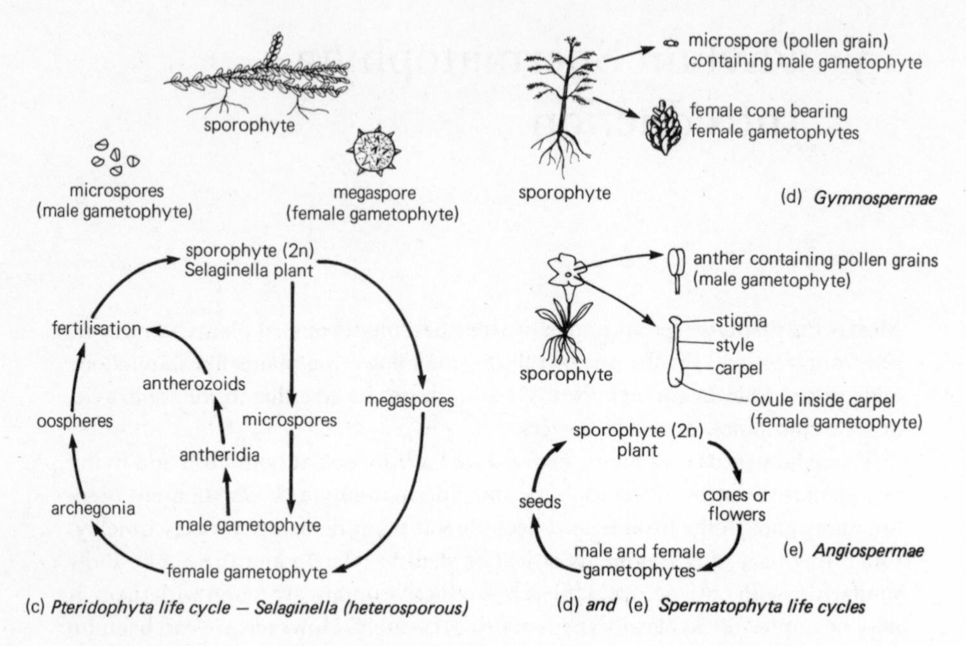

Figure 75(cont.) Comparison of plant life cycles, (c), (d) and (e)

Some of the Pteridophyta have developed two kinds of spores in the sporophyte generation (see *Selaginella* p. 125). In this heterosporous condition the microspores, which contain the male gametes, are smaller than the megaspores, which contain the female gametes. This kind of life cycle can be seen in Figure 75c. Both kinds of spore are ultimately released from the sporophyte and develop into separate gametophytes, but the megaspores often begin their development while still inside the megasporangium.

The Spermatophyta are also heterosporous but they do not develop an independent gametophyte generation. The microspores are pollen grains which may be carried by wind or insect pollination to the vicinity of the megaspores. These develop on the scales of a female cone or enclosed inside a flower. See Figure 75d and e. The megaspore remains enclosed inside the ovule or megasporangium throughout the processes of pollination and fertilisation. After fertilisation the ovule is called a seed and it is only at this stage that it is finally set free from the sporophyte plant.

CLASSIFICATION

The Spermatophyta are divided into two classes, the Gymnospermae which are cone-bearing and the Angiospermae which bear flowers. The Gymnospermae have their ovules exposed on the scales or megasporophylls of a female cone. The microspores or pollen grains are wind-borne and are blown into the female cones. There they may come to lodge inside a small opening to the ovule, called the *micropyle*. The pollen grains then develop *pollen tubes* down into the female gametophyte to the region of the archegonia. One order of the Gymnospermae, the Cycadales, are probably among the most primitive living gymnosperms. Members of this order, which includes *Cycas*, the sago palm, have male gametes which swim to the archegonia. This is the *zoidogamous* condition; whereas the non-motile male gametes of the rest of the gymnosperms and all the angiosperms, represent the *siphonogamous* condition.

The angiosperms have their ovules enclosed inside one or more *carpels*, each of which represents a megasporophyll which has become folded over and fused along its edges, forming a container. The upper part of the carpel develops into a *style* and a *stigma* (Figure 75e) on which pollen is deposited. After pollination a pollen tube grows down from the pollen grain through the style to the ovule inside the carpel. The tip of the tube grows towards the micropyle of an ovule. The male gamete has been carried down inside the tube as it grew, and now emerges and eventually fuses with the female gamete inside the ovule. This process is described in more detail in chapter 17.

After fertilisation in an angiosperm the ovules become *seeds* and the carpel(s) becomes the *fruit*. Apart from the obvious advantage of protection, the carpels may also be modified to aid in dispersal of the seeds, see p. 150.

In both the gymnosperms and the angiosperms the seeds may be dormant for some time after their release from the parent plant, before they germinate into a new sporophyte. The viability of their seeds, often over a period of years, is a significant advantage for a terrestrial existence.

The Spermatophyta include a great variety of habit, from small herbaceous plants to large woody shrubs and trees. *Herbaceous* plants are composed only of soft tissues whereas woody ones contain cells with much more rigid walls. In some cases the older roots and/or stems may continue to increase in diameter as the plant ages. This is the result of new cells being added by an actively dividing layer called the *cambium*. This kind of growth is called *secondary thickening*. A further cambium usually develops as the stem or root enlarges and this produces a protective layer of *cork* or *bark* around the outside. Such woody plants may be called *shrubs* or *trees*. A tree has a single main trunk whereas a shrub has a number of trunks of similar size.

16 Class Gymnospermae

The gymnosperms appear as fossils during the Upper Devonian period about 300 million years ago. They were the dominant form of vegetation during the Jurassic and Cretaceous periods and since then have been largely replaced by the angiosperms.

Their most characteristic feature is that their seeds are exposed, not enclosed in an ovary. However there seems to be an evolutionary trend in the gymnosperms towards greater protection of the ovules, by the development of enclosing integuments, by grouping them together in cones, or within red juicy cups or *arils*, such as those of the yew, which also attract birds and help in seed distribution.

The gymnosperm sporophyte is a woody plant with true roots, stems and leaves. They show great range of size, from the shrubby, prostrate junipers to the giant Californian redwood trees. Most gymnosperms are evergreen and have pointed or scale-like leaves. The conifers in particular are economically important in supplying timber, resin, turpentine and wood pulp.

CLASSIFICATION

The Gymnospermae number about 700 species, of which about 600 belong to the Order Coniferales. They include well-known cone-bearing plants like pine, larch and spruce. The larches are deciduous, but the majority of conifers are evergreen. They are found particularly in temperate northern regions, but also on cooler mountain slopes within the tropics. Their small leaves help them to withstand drought on poor soil or in freezing conditions, by providing only a very small surface area for evaporation of water.

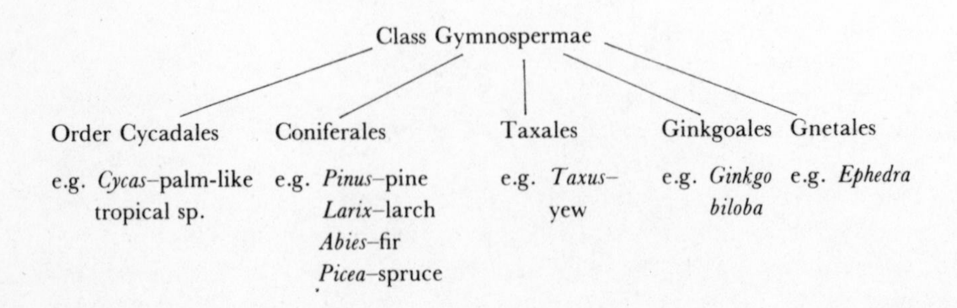

The Order Ginkgoales contains a single species, *Ginkgo biloba*, the maidenhair tree, which has fully exposed seeds. The Order Gnetales contains three genera which show the greatest similarity to the angiosperms, but are all very different from each other. Unfortunately there are no fossil remains to help trace their evolutionary relationships.

Pinus is described as a typical gymnosperm.

Pinus sylvestris (the Scots pine)

The Scots pine is a tree common all over Britain, even on rocky or sandy soil. It has a deep tap root, or more spreading roots if the soil is shallow, and a trunk covered in rough reddish-brown bark. The main branches of the tree increase their length each year by the activity of the terminal bud at the tip and are called *branches of unlimited growth*. At intervals are groups of lateral buds which develop into whorls of lateral branches. A young tree has a symmetrical appearance which is lost as it matures because the lower branches tend to fall off, leaving a tall bare trunk, (Figure 76a).

The branches are covered by spirally-arranged scale leaves. The needle-like foliage leaves are on very short *branches of limited growth* called *dwarf shoots*, which develop in the axils of the scale leaves. In *P. sylvestris* there are two leaves on each dwarf shoot. They stay on the tree for two to three years, then the dwarf shoot and the leaves fall off together, (Figure 76b and c).

The tree is the sporophyte generation of *Pinus*. It is heterosporous, producing two kinds of spores, microspores and megaspores. These develop on sporophylls arranged as cones. Both male and female cones are produced on the same tree and they begin their development in spring. In May groups of young male cones appear in the axils of scale leaves near the bases of new shoots. Each male (staminate) cone consists of a central axis covered by up to one hundred spirally-arranged microsporophylls. On the underside of each microsporophyll are two microsporangia (pollen sacs), filled with microspores (pollen grains) (Figure 76d). Each microspore has two lateral air sacs which will aid its dispersal by wind (Figure 76e). The microspores have been produced by meiotic division from microspore mother cells and are therefore haploid. They are released when the walls of the microsporangia split longitudinally.

The male cones differ from the flower of an angiosperm in the following ways: a) the axis (receptacle) is elongated, b) only the male organs are present, c) there is no division into an anther and filament region, d) there are only two pollen sacs instead of four.

The female cones begin their development at about the same time as the male cones, but they take three years to produce fertile seeds. Female cones arise in the axils of scale leaves on lateral branches away from the male cones and usually only one or two develop on a branch (Figure 76b). The young cones are green but they become brown with age. Each cone consists of spirally-arranged outgrowths around a central axis. The outgrowths are made up of a leathery *bract scale* growing from the axis and a larger, woody *ovuliferous scale* growing from its upper surface (Figure 76f). As the ovuliferous scale enlarges the bract scale

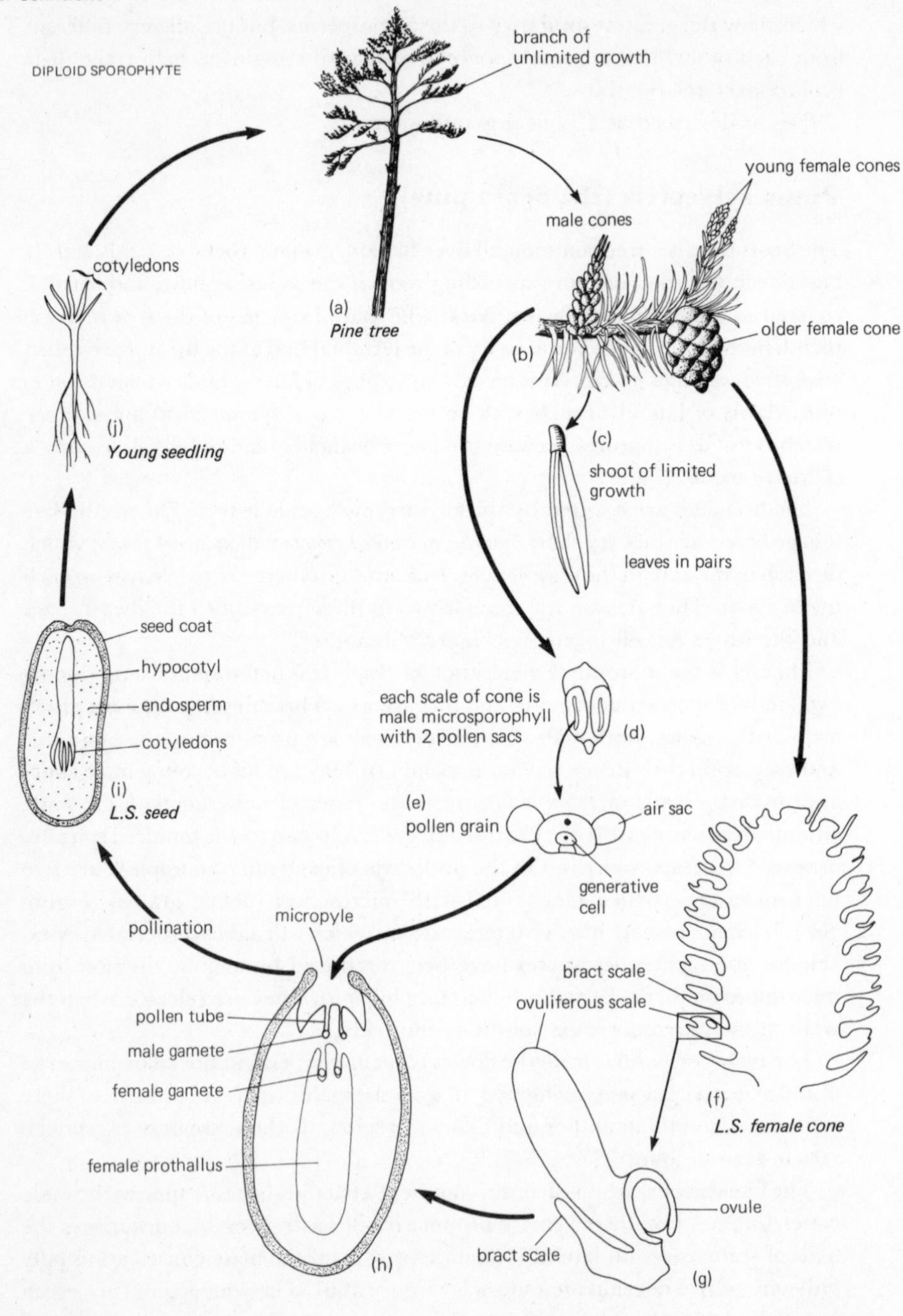

Figure 76 Life cycle of *Pinus*

becomes hidden, though it remains visible in some conifers such as larch. On the upper surface of the ovuliferous scale are two ovules lying side by side. The ovules develop from surface cells of the scale which then become enclosed by upgrowths of the surrounding tissue. The inner region becomes the *nucellus* and the upgrowths form the *integument*. This covers the nucellus except for a tiny pore at one end, called the *micropyle*.

Near the micropyle end of the nucellus is a large *archesporial cell*. It divides into two cells, a single *tapetal cell* and a megaspore mother cell. The latter divides by meiosis forming four megaspores, but only one of these matures.

The female cone may be regarded as a single flower with the bract scales as open carpels bearing ovules, or the cone axis can be regarded as a branch and each ovuliferous scale as a female flower, so that the whole structure is an inflorescence.

The microspore begins its development before being shed in May. The protoplast divides into a small prothallial cell and a larger tube cell. No further development occurs until after the pollen has been shed and has blown in through the micropyle of a megaspore. Then the tube cell divides again producing a second small *prothallial cell*. The prothallial cells have no function, they are the only vegetative part of the male gametophyte and soon disappear. The tube cell divides again forming an antheridial cell which rests until the following spring. The tube cell meanwhile grows a tube down into the nucellus (Figure 76h). Once the microspores have been shed the male cones wither and fall off. The scales of the female cones open at pollination time and close up again later. The ovules also secrete mucilage through the micropyle. This tends to trap microspores which are then drawn inwards as the mucilage dries.

Throughout the pollination period the megaspores have continued their development. The megaspore enlarges and is now called an *embryo sac*. Its nucleus divides many times and the embryo sac becomes cellular. This tissue represents the female gametophyte. Unlike the megaspore of *Selaginella*, that of *Pinus* continues its development inside the megasporangium. At the micropyle end two or three archegonia develop. Each consists of a short neck without neck canal cells, and a venter with an oosphere and ventral canal cell. Archegonial development is essentially the same as that of the fern or *Selaginella*. The female cones overwinter in this condition.

The following April, inside the microspores the antheridial cell divides again. This division results in an antheridium and a stalk cell. The antheridium divides yet again into two male gametes. The pollen tube continues to penetrate the female gametophyte and eventually reaches the neck of an archegonium. It grows down into the venter and the end of the tube then ruptures. All the contents of the tube disintegrate except for one male gamete, which fertilises the oosphere. If more than one oosphere is fertilised only one continues to develop.

Fertilisation is usually completed by June of the year after pollination. The fertilised oosphere or oospore begins its development into an embryo while still inside the ovule on the parent plant (Figure 77). The nucleus of the zygote divides mitotically three times forming eight nuclei arranged in two tiers of four. The upper four take no further part in the development. The lower four become

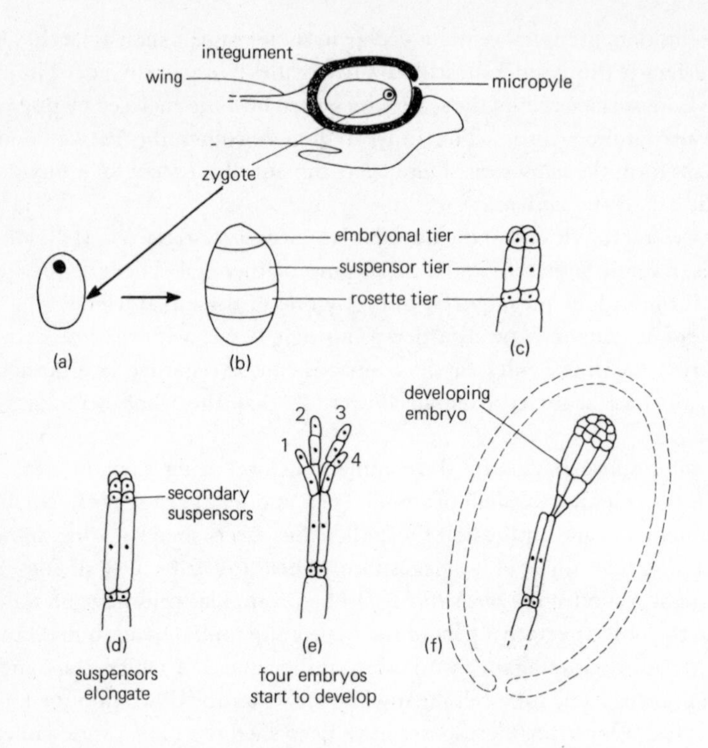

Figure 77 Stages in the development of *Pinus* embryo

surrounded by cell walls and then each divides twice resulting in three tiers of four cells (Figure 77c). The top four cells are the rosette cells, the middle ones are the suspensors and the lowest ones are the embryonal cells. The suspensor cells elongate, pushing the embryonal cells deep into the cells of the female gametophyte (Figure 77d). Each embryonal cell also elongates and divides into two cells, a secondary suspensor and an embryo. Thus four embryos begin to develop, a feature called *polyembryony*, which is characteristic of conifers. However, only one normally goes on to maturity (Figure 77e and f).

The embryo differentiates into a radicle at the micropyle end and a plumule surrounded by five to ten cotyledons at the other end. Food for the growth of the embryo comes from the female gametophyte, and any unused when the embryo is complete remains as a food store. This is referred to as *endosperm* but is not homologous with the endosperm of an angiosperm because it is part of the female gametophyte and not formed by the fusion of gametes (Figure 76i).

As the embryo matures the integument changes into a hard covering – the *testa*, and the seed is complete. Meanwhile the ovuliferous scales of the female cone have become brown and woody. When the seeds are fully formed the scales bend outwards exposing the two seeds on each of them. The seeds are winged and are soon shaken out and blown away.

If conditions are suitable where the seed lands, germination soon begins. The testa splits and the radicle emerges and grows down into the soil. The cotyledons are carried above ground with the plumule so that germination is *epigeal* (Figure 76j).

The production of a *Pinus* seed has taken three growing seasons and the events may be summarised as follows:

Year 1 male and female cones develop and produce microspores and megaspores, pollination, gametophytes develop;

Year 2 fertilisation, embryo begins to form;

Year 3 seeds mature and disperse.

Adaptations for a terrestrial existence

Pinus has a well-developed root system for anchorage and absorption, and mechanical tissue supporting a thick, tall stem. The tree habit allows easy access to light and air and aids dispersal of spores and seeds, but there are also disadvantages. Over-wintering above ground means exposure to frost and perhaps little available water and although the leaves are retained through the winter they do show several adaptations for avoiding water loss, such as their small surface area and thick cuticle.

Trees have a long reproductive life but pines are slow to mature and must survive many years before they are able to reproduce. In comparison with the Bryophyta and Pteridophyta the *Pinus* life cycle shows some advances. The microspores are no longer dependent on a film of water for their dispersal, but are spread by wind. Fertilisation too avoids the use of swimming gametes by the development of a pollen tube similar to that of the angiosperms.

QUESTIONS

1 By means of life cycle diagrams compare the position and relative importance of the gametophyte and sporophyte generations in the life cycles of a bryophyte, a pteridophyte and a gymnosperm.

2 *Pinus* is better adapted than a fern for life on land. Discuss.

3 Describe the changes which occur as the fertilised ovule of *Pinus* develops into a mature seed.

17 Class Angiospermae

The angiosperms or flowering plants appear in the fossil record during the Cretaceous period, 135 million to 70 million years ago. Since this period they have become the dominant form of vegetation on land throughout the world. There are about 300000 species of angiosperms – more than all the other plant species put together. They provide vital food sources for mankind in the form of cereals, root crops, leaf crops and fruits. Other important products include textiles, drugs, vegetable oils, cordage, timber and fuel.

THE DIVERSITY OF THE FLOWERING PLANTS

Angiosperms exhibit a great range of shapes, sizes and modifications by which they are able to colonise habitats on land, in freshwater and even to some extent salt water: *hydrophytes* live in water or very wet places and include pond weeds, water lilies, sedges and rushes; *mesophytes* are found in the usual water conditions of grassland, hedges and woods; *xerophytes* are able to withstand drought, either by being able to survive partial desiccation, as in the spurges and other desert shrubs, or by storing water, as cacti do in their stems; *xeromorphs* are structurally similar to xerophytes yet die if exposed to drought. They include *halophytes*, salt marsh plants, able to absorb water and lose it by transpiration even when subjected to salt spray and sea water.

Flowering plants may be either woody trees or shrubs or herbaceous plants – composed of soft tissues only. Many woody species are *deciduous*, losing all their leaves at the approach of winter and regaining them in spring e.g. oak, beech. Others, such as holly and privet retain their leaves through the cold season and lose them gradually throughout the year.

Non-woody plants may also be perennial, surviving the winter mainly as underground structures, protected from bad weather e.g. bulbs, corms, rhizomes etc.

During the cold season the aerial parts of the plant wither away, to be replaced by new shoots in the spring. A few, such as dandelions and most grasses, retain some of their leaves through the winter and are called *winter-green perennials*.

Many herbaceous plants are *annuals*, completing their life cycle from germination to seed production in a single year. A few, such as groundsel, are *ephemeral plants* having such a short life cycle that more than one generation can be produced in one season. A *biennial* requires two seasons to reach maturity. In the first season it produces a rosette of leaves and a store of food in the undergound parts of the plant. It overwinters in this condition and uses the stored food to produce rapid growth and flowering during the second season. Wallflowers and carrots are examples of biennials.

CLASSIFICATION

The flowering plants are divided into two large sub-classes as shown below:

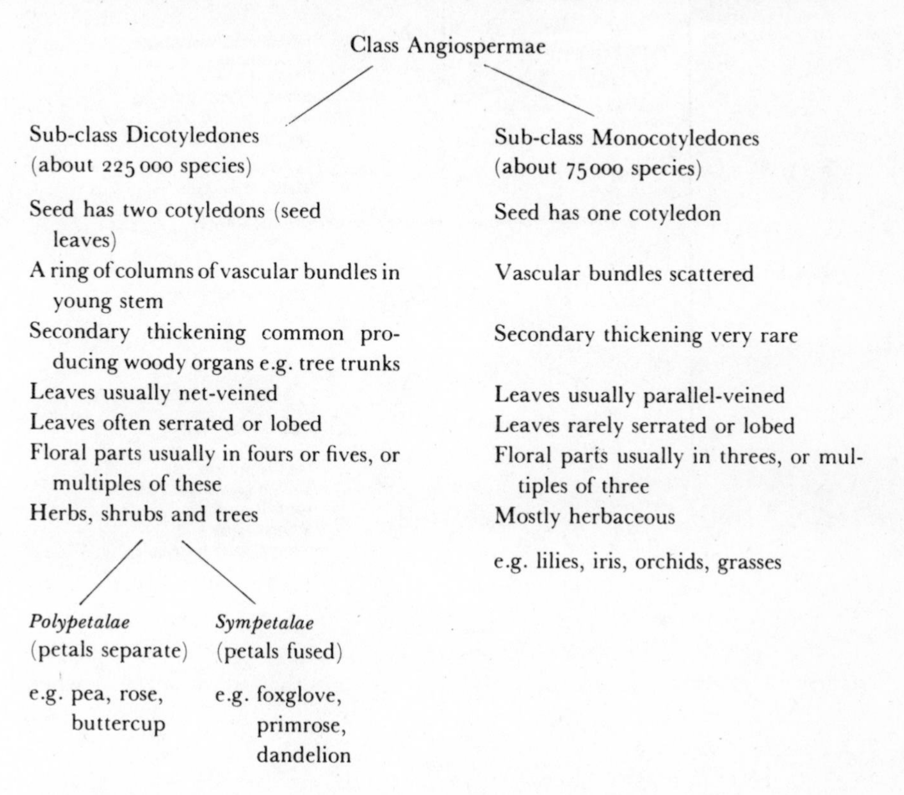

Class Angiospermae

Sub-class Dicotyledones (about 225 000 species)	Sub-class Monocotyledones (about 75 000 species)
Seed has two cotyledons (seed leaves)	Seed has one cotyledon
A ring of columns of vascular bundles in young stem	Vascular bundles scattered
Secondary thickening common producing woody organs e.g. tree trunks	Secondary thickening very rare
Leaves usually net-veined	Leaves usually parallel-veined
Leaves often serrated or lobed	Leaves rarely serrated or lobed
Floral parts usually in fours or fives, or multiples of these	Floral parts usually in threes, or multiples of three
Herbs, shrubs and trees	Mostly herbaceous
	e.g. lilies, iris, orchids, grasses

Polypetalae (petals separate)	*Sympetalae* (petals fused)
e.g. pea, rose, buttercup	e.g. foxglove, primrose, dandelion

The dicotyledons are divided into groups with the petals separate from each other, the Polypetalae, or fused, the Sympetalae. The Polypetalae are considered to be the more primitive group.

Here are some examples of the range of angiosperm habit:
Herbaceous dicotyledon – Wallflower (Figure 78);
Herbaceous monocotyledon – Meadow fescue (Figure 79);
Woody dicotyledon a) evergreen shrub – Rhododendron (Figure 80);
b) tree – Horse chestnut (Figure 81).

THE STRUCTURE OF A FLOWER

A flower is complete if it possesses the full complement of floral parts (Figure 82). At the base of the flower is the *receptacle* and on it are borne the parts of the flower.

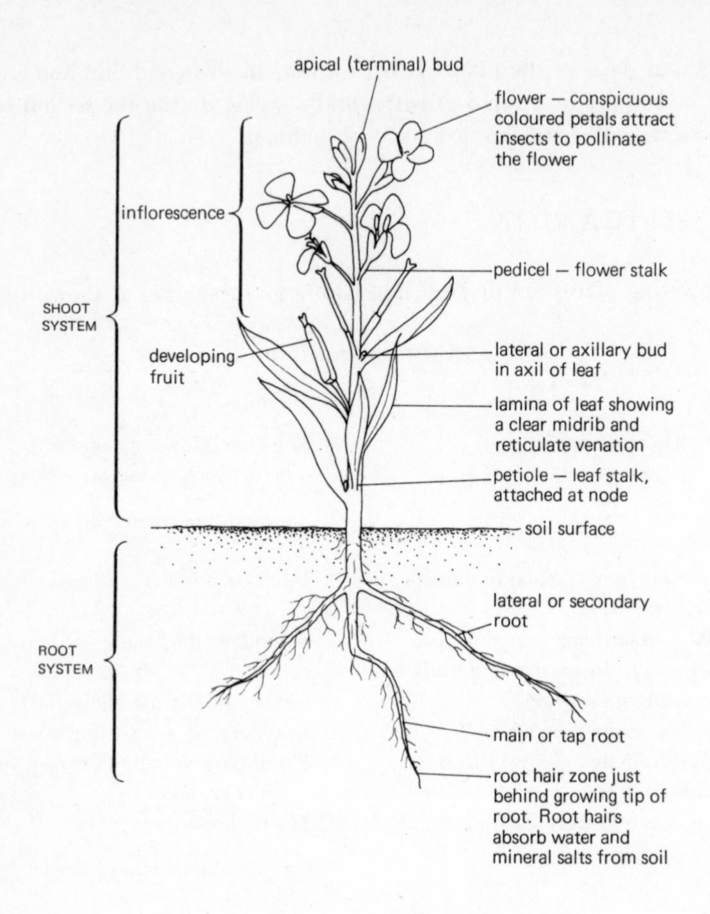

Figure 78 The Wallflower, *Cheiranthus cheiri*, is a typical herbaceous dicotyledon growing to about 30 cm. The *lamina* or leaf blade has a network of veins – *reticulate venation*. The points of leaf attachment are called *nodes* and the spaces between them *internodes*. These become shorter towards the *apical bud* at the top. Lower down are *axillary buds* produced in the axils of the leaves (in the angle between the leaf and the stem). The flowers are borne at the top of the shoot in a group called an *inflorescence*.

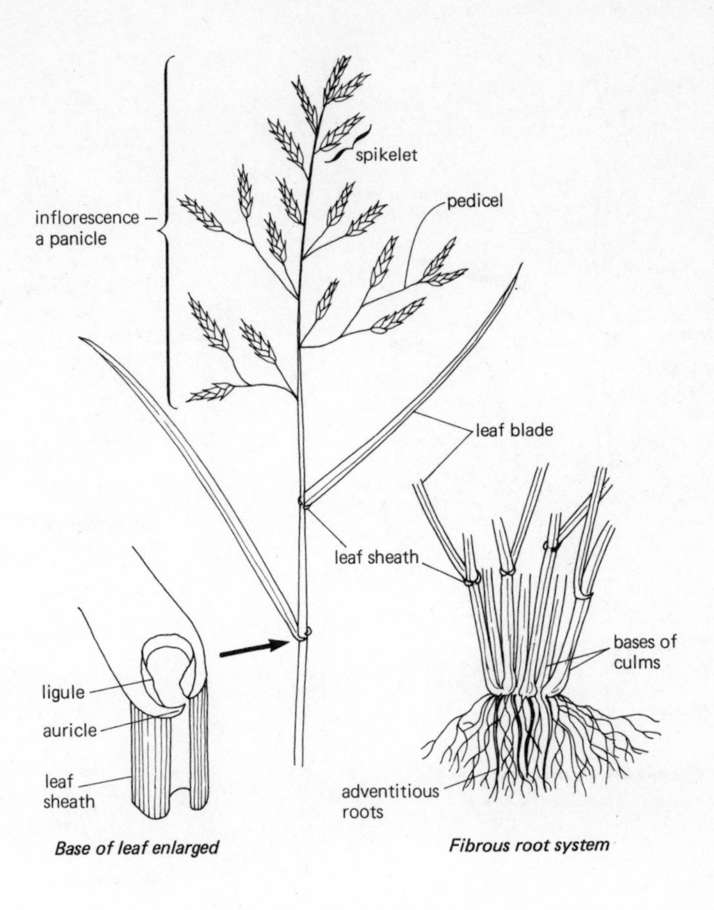

Figure 79 A herbaceous monocotyledon – Meadow fescue
Meadow fescue, *Festuca pratensis*, is a typical perennial grass plant varying in height from 30 to 120 cm. The root system is *fibrous* – having no main root, but developing *adventitiously* from the bases of the stems. The shoot system is made up of a number of erect, unbranched stems or *culms*. There are two to four nodes where leaves are given off, alternately in two rows down the stem. The upper part of each leaf is a long tapering blade about 45 cm long and 3 to 8 mm wide. The lower part is a *sheath* which curves around the stem. Where the sheath joins the blade there is a thin membranous outgrowth called a *ligule* and two tiny pointed projections – the *auricles*.

The flower head is borne at the top of the culm. It consists of a main axis 10 to 35 cm long, bearing a number of *spikelets* attached by short stalks or pedicels. Each spikelet is composed of five to fourteen flowers. The whole inflorescence is called a *panicle*.

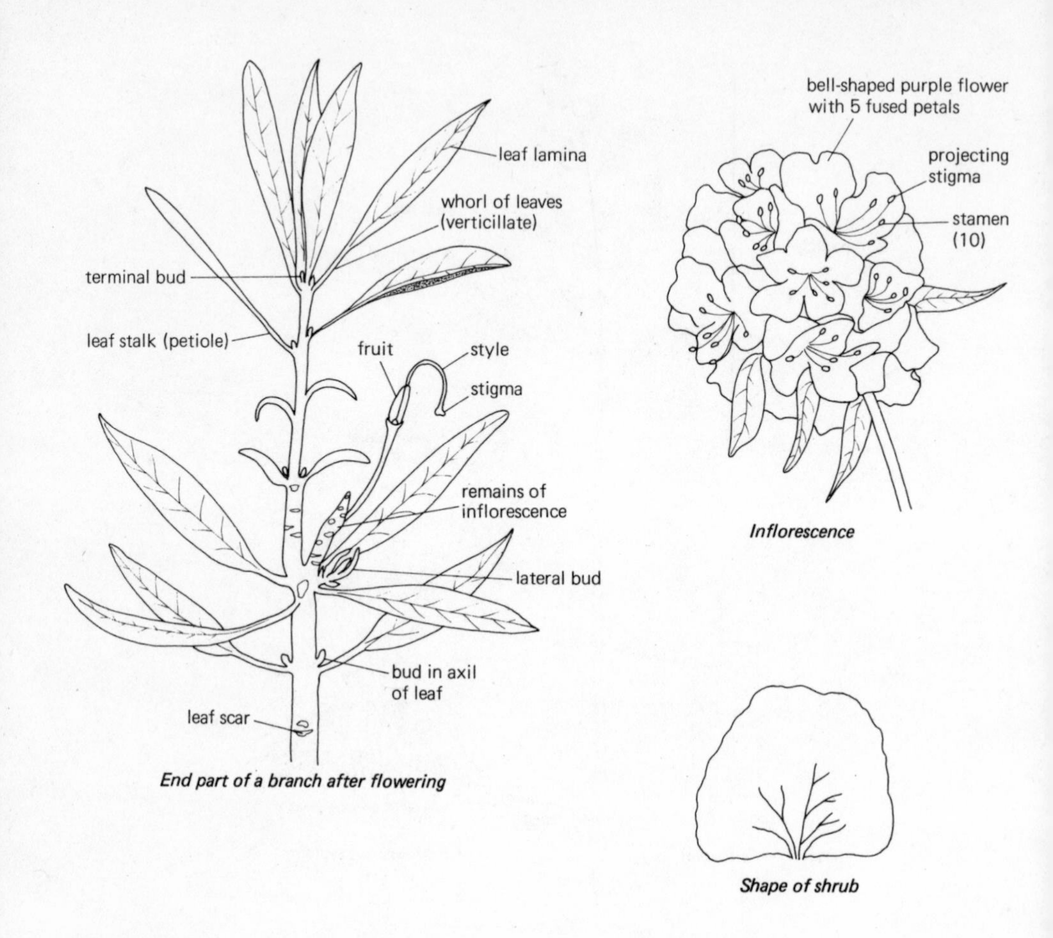

Figure 80 *Rhododendron ponticum* is a large evergreen garden shrub, now naturalised on heaths and in woods on sandy or peaty soil. It develops many stems of similar size which later become woody and may reach a height of several metres. The leaves are long and pointed, becoming darker and shinier with age. They are attached alternately along the stems and as a whorl around the terminal bud – a *verticillate* arrangement. Inflorescences of large, purple, bell-shaped flowers appear in May–June.

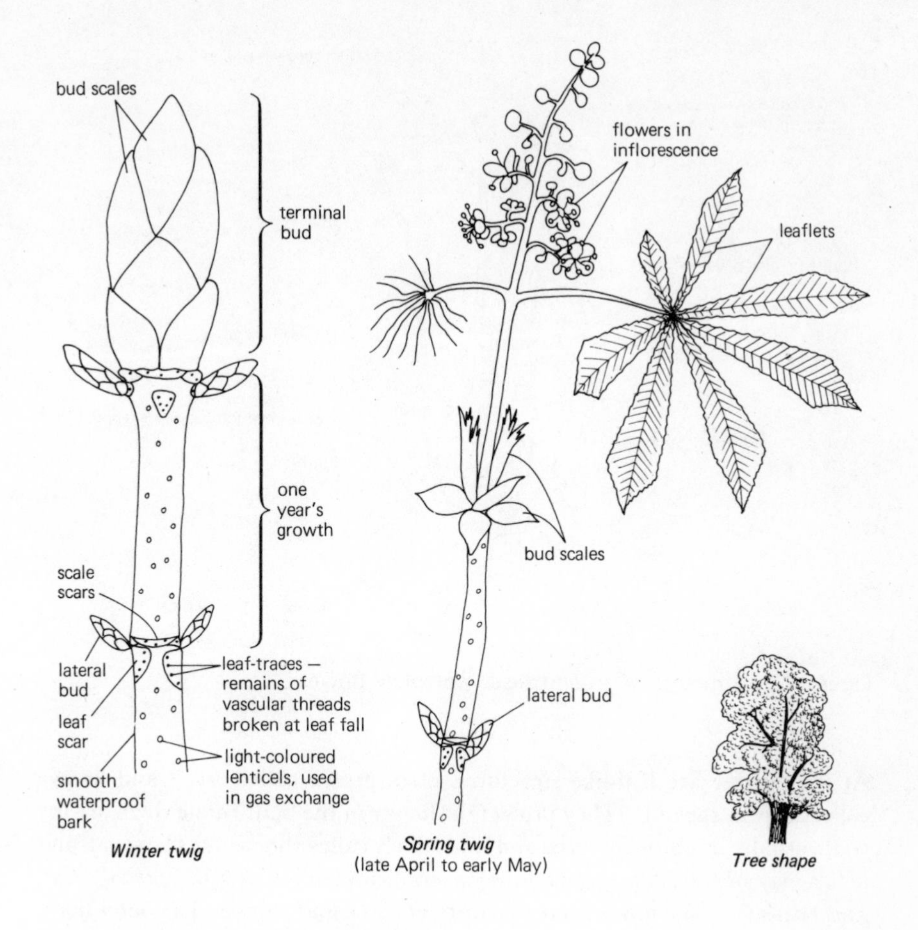

Figure 81 A woody dicotyledon – horse chestnut

The Horse chestnut, *Aesculus hippocastanum*, is a tall deciduous tree, reaching 30 m or more in height. It has an extensive root system with the older roots becoming woody as a result of secondary thickening. The single stem also becomes woody, forming a stout trunk. A *cork cambium*, a layer of actively dividing cells, develops near the outside of the stem and begins to lay down waterproof *bark* around the outside of the stem as it increases its girth.

The branches bear large buds, in pairs opposite each other and covered at first in sticky brown, waterproof *scales* which protect the foliage leaves inside. When the buds open the scales fall, leaving a ring of small scars. The distance between one group of scars and the next represents one year's growth of the twig.

The leaves have five to seven leaflets attached to the tip of a long stalk. In autumn the leaflets are shed separately and the fall of the leaf stalk leaves a triangular scar.

The mature tree bears flowers in an inflorescence produced by the terminal bud, so that further growth of the branch is by the lateral buds.

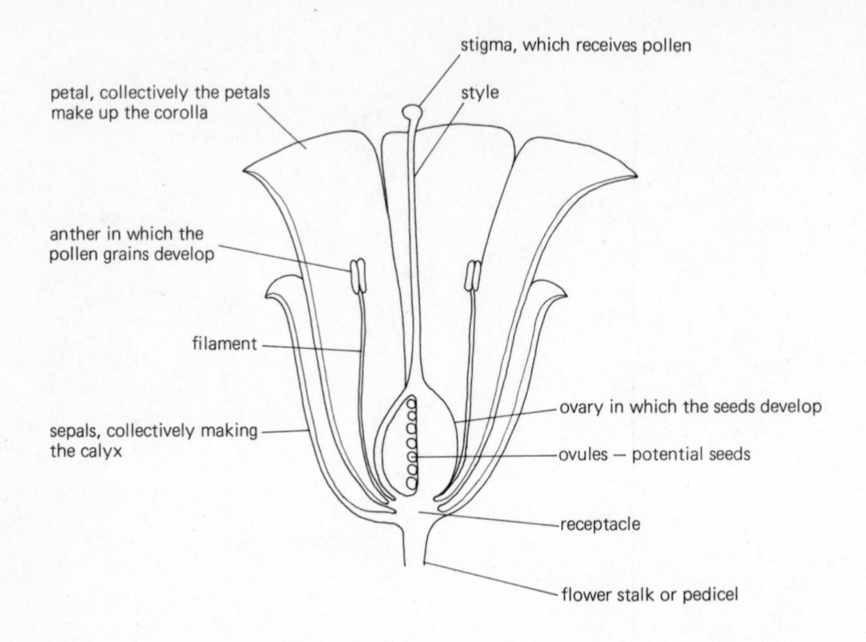

Figure 82 Structure of a generalised complete flower

At the outside are leaf-like structures, often green, called *sepals* and known collectively as the *calyx*. They protect the flower in the bud. Inside them are the *petals*, usually brightly-coloured and collectively called the *corolla*. The sepals and petals may be free (*polysepalous, polypetalous*) or united into a tube (*gamosepalous, gamopelalous*). Some flowers have no distinct calyx and corolla. The outer floral parts are then called the *perianth* e.g. Bluebell.

Inside the corolla lie the *stamens*. Each consists of a *filament* supporting an *anther*, in which the pollen grains (containing the male gametes) are produced. Collectively the stamens form the *androecium*. At maturity the anthers split open and the pollen is exposed. If the anthers split down their outer sides it is *extrorse dehiscence;* if the split occurs down the inside the dehiscence is *introrse*.

In the centre of the flower is the *gynaecium* or *pistil*. It is composed of one or more *carpels* containing *ovules*, in which are the female gametes. If a number of carpels are present they may be separate (*apocarpous*) or united to form a single ovary (*syncarpous*). Attached to the ovary is a *style* with a *stigma* at its tip. The stigma receives pollen during pollination.

A detailed description of a flower requires investigation into the arrangement of the ovules inside the ovary, (Figure 83). A syncarpous ovary formed by the joining of a number of carpels may retain a number of internal cavities, often one per carpel. These cavities are called *loculi* (sing. *loculus*) and may be seen if the ovary is cut across transversely. In some cases when the carpels fuse the internal walls separating them are lost so that a single loculus is formed.

The ovules are attached to the ovary wall by a *placenta* and their arrangement is referred to as their *placentation*. Figure 83 shows the main types of placentation.

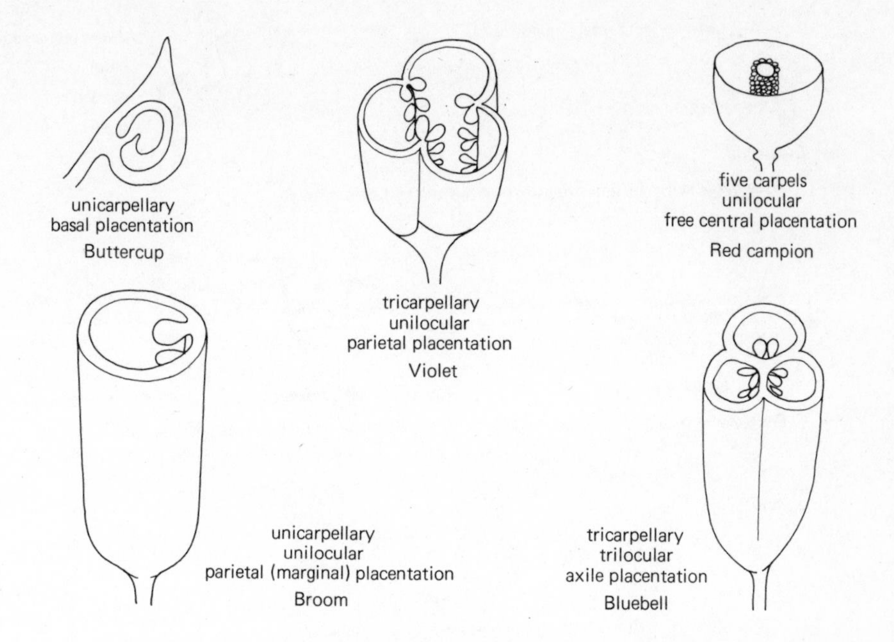

Figure 83 Arrangement of the ovules in the ovary

Not all flowers are complete. Grass flowers for example, lack a colourful corolla. Most flowers are hermaphrodite – having both androecium and gynaecium, but some species have separate *staminate* and *carpellary* flowers i.e. the flowers are unisexual. Oak and hazel have male and female flowers on the same plant and are *monoecious*; whereas willow and poplar have male and female flowers on separate plants. This is the *dioecious* condition.

The symmetry of a flower

Radially symmetrical flowers are called *actinomorphic* or *regular*; whereas bilaterally symmetrical flowers are *zygomorphic* or *irregular*.

The terms anterior, posterior and lateral are often used to describe the relative positions of the parts of a flower. They can only be used to describe a flower produced as a lateral bud and cannot be applied to a terminal solitary flower. The *posterior* of the flower is the side nearest the stem; the *anterior* is the side facing away from the stem.

Insertion of the floral parts on the receptacle

The relative position of the floral parts depends on the shape of the receptacle and may be *hypogynous*, *perigynous* or *epigynous*, (Figure 84). The gynaecium is *superior* if it is borne above the rest of the flower, or *inferior* if the sepals, petals and stamens are inserted above it.

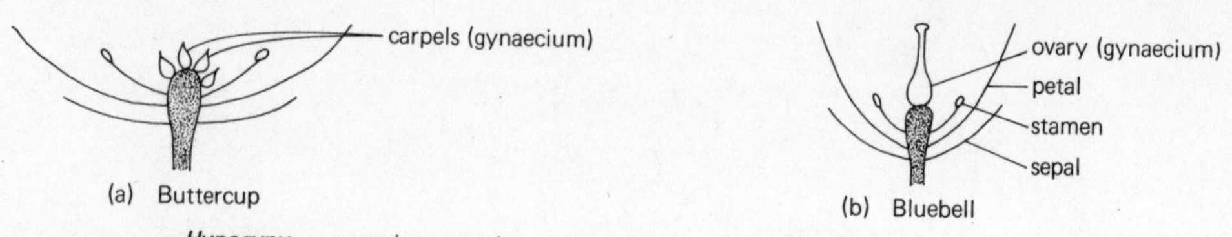

Hypogyny – gynaecium superior – above the rest of the flower

Perigyny – receptacle saucer-shaped, floral parts appear to be level with gynaecium

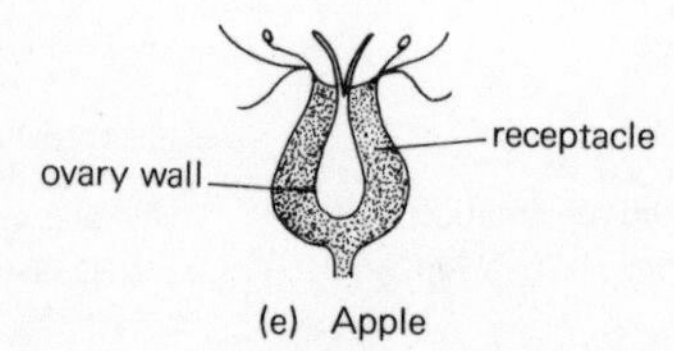

Epigyny – receptacle a deep cup, fused to ovary wall, gynaecium inferior

Figure 84 Insertion of the floral parts on the receptacle

THE LIFE CYCLE OF A FLOWERING PLANT

Pollination

Inside the anthers *pollen grain mother cells* produce haploid *pollen grains* (microspores) by meiosis. The pollen grains are liberated by the dehiscence of the anthers and are usually carried away either by wind or insects. The transmission

Table 2

Insect-pollinated (entomophilous)	*Wind-pollinated (anemophilous)*
1 Flowers brightly-coloured, often scented and/or with nectar. Small flowers arranged in clusters.	Flowers small, inconspicuous, no scent or nectar.
2 Pollen grains sticky usually sculptured – adhere to insect.	Small smooth light pollen grains, may be blown some distance.
3 Stigma(s) inside flower.	Large, feathery stigma(s) hang outside flower.

of the pollen by these agencies from the anthers to a receptive stigma is *pollination*. If the pollen travels to the stigma of the same flower or another flower on the same plant it is *self-pollination*, but if it is carried to a flower on another plant of the same species it is *cross-pollination*.

Wind-pollinated flowers include the grasses and many trees such as oak and hazel, where the height above ground may aid wind dispersal. Some anemophilous species also produce their flowers before the foliage, leaving them more exposed to wind.

Pollination may also occur by water, as in the Common grass-wrack *Zostera marina*, where the pollen floats through the water to submerged feathery stigmas.

Cross-pollination

There are various methods whereby plants may avoid self-pollination. Some, such as many cultivated forms of apple and pear are *self-sterile* and if pollen lands on the stigmas of the plant that produced it, it will not germinate. Another method is to separate the anthers and stigmas, either by space or by the time of their development. Unisexual flowers produced on separate plants as they are in the campions obviously make self-pollination impossible. It is quite common for the anthers to mature before the gynaecium, a condition known as *protandry* and found in dandelion, foxglove, monkshood and many others. Much less common is *protogyny*, when the gynaecium develops before the anthers are mature. Examples of this include *Scrophularia* (figwort), bluebell and crab apple.

Special pollination mechanisms

A few flowers show particular arrangements of their floral parts to make cross-pollination likely (Figure 85). For example the primrose produces flowers of two kinds – (dimorphism) – known as *thrum-eyed* and *pin-eyed*. In the pin-eyed form the stigma is near the top of the flower and visible like a pin's head, whereas in the thrum-eyed form the anthers are near the top of the flower and the stigma is lower down. The difference is produced by having styles of differing length and termed *heterostyly*. If a bee lands on a thrum-eyed flower and pushes its proboscis down to the bottom for nectar, its head touches the anthers and pollen sticks to it. This part of its head is in exactly the right place to touch the stigma of a pin-eyed flower. Similarly, on visiting the pin-eyed flower, pollen is received in the correct position for touching the stigma of a thrum-eyed flower.

A particularly delicate floral mechanism is shown by the sage flower, *Salvia*. It has two stamens, each of which has a fertile upper lobe producing pollen and a lower sterile lobe which does not. The style is long and holds the stigma up above the anthers. An insect landing on the lower lip of the corolla, pushes its head down the corolla tube to reach the nectar. In so doing it pushes against the sterile lower lobe of the anthers and makes the upper pollen-producing lobes swing down to touch its body. Later, after the stamens have withered the style elongates and curves down to bring the stigma into a position where it will touch the back of any visiting insect and receive any pollen that it carries.

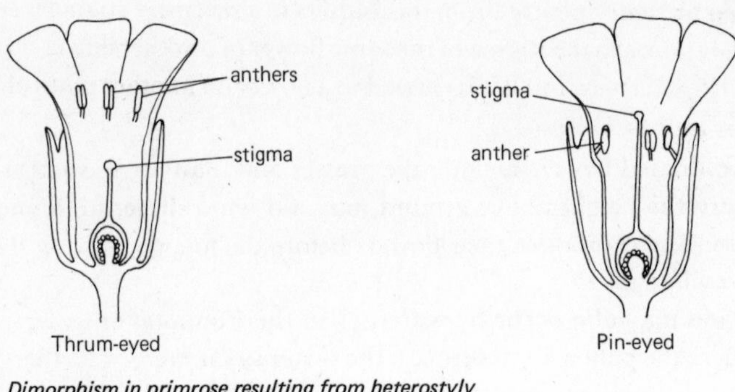

Dimorphism in primrose resulting from heterostyly

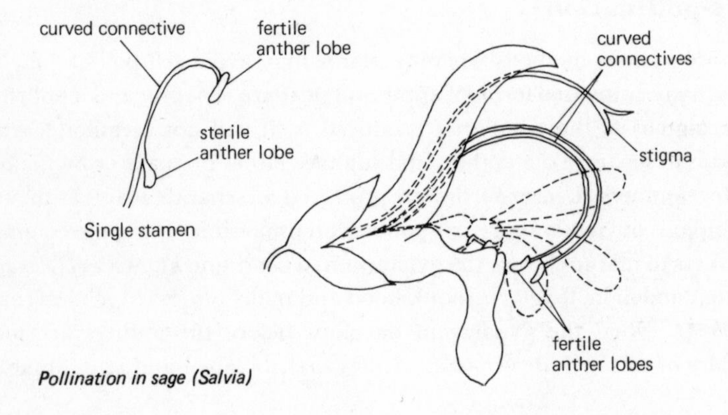

Pollination in sage (Salvia)

Figure 85 Special pollination mechanisms

Self-pollination

Many annuals, such as groundsel and chickweed are regularly self-pollinated. They produce no nectar or scent, the anthers and stigmas mature at the same time and are close together in the flower. Violet and wood sorrel produce some *cleistogamic buds* which never fully open. Inside them pollen falls from the anthers to the stigma and ensures that at least some flowers are pollinated. If cross-pollination fails to occur flowers such as the dandelion allow the anthers and stigma to touch, resulting in self-pollination.

Despite the common occurrence of self-pollination in many flowers, cross-pollination would seem to offer definite advantages. It avoids the propagation of deficiencies and weaknesses from one generation to the next thus leading to more numerous and stronger offspring and allows greater possibilities of variation.

The male and female gametophytes

When mature the pollen grain, which is the male gametophyte, contains a *tube nucleus* and a *generative nucleus*. If it comes to rest on a receptive stigma it will

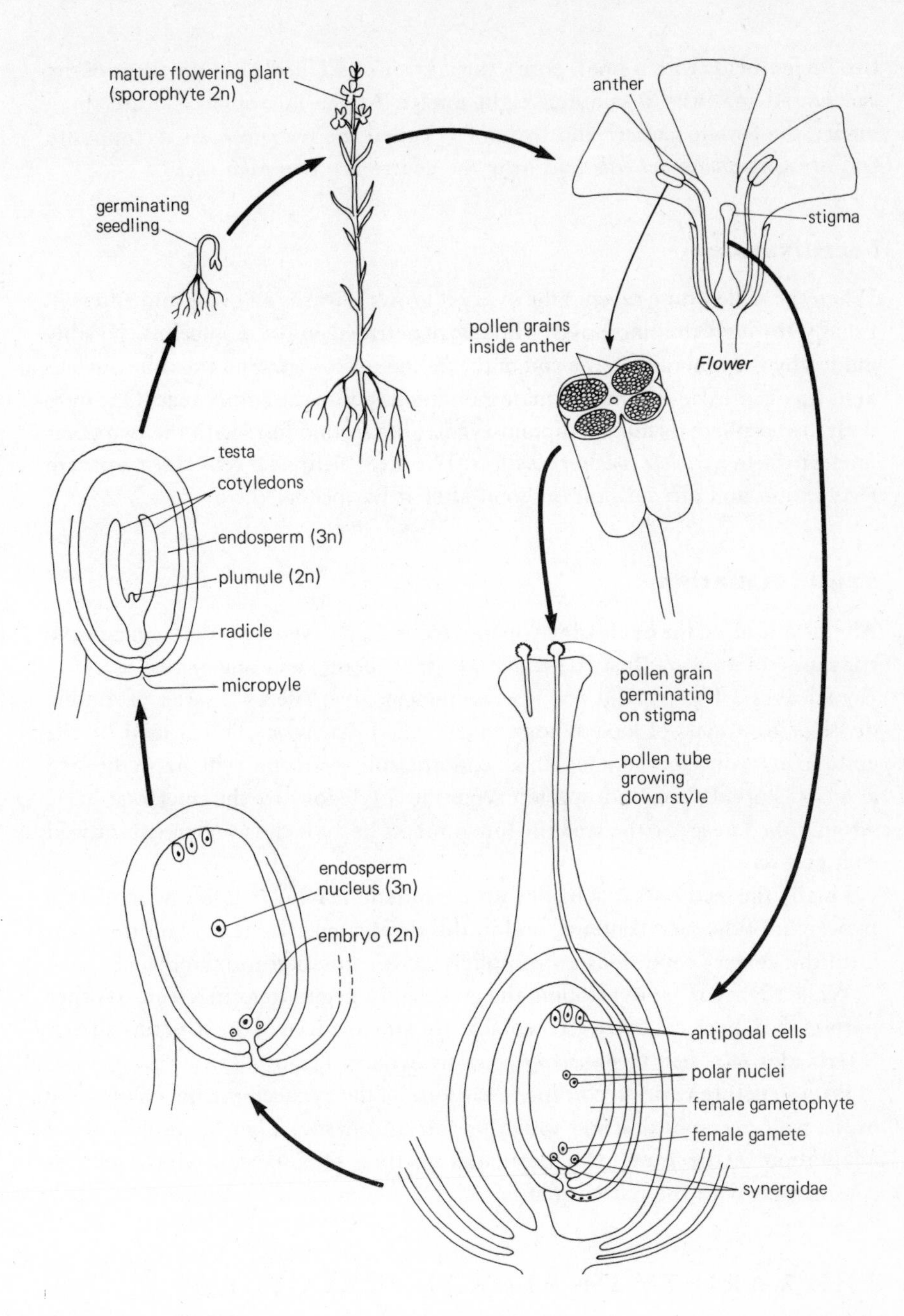

Figure 86 Life cycle of a flowering plant

germinate and develop a pollen tube which grows down the style to the ovary (Figure 86). Growth of the pollen tube is controlled by the tube nucleus. During this period the generative nucleus divides to produce two male gametes.

The female gametophyte is the *embryo sac* inside the ovule. Surrounding it are two integuments with a small pore – the *micropyle* and inside it is the tissue of the *nucellus*. At maturity it contains eight nuclei. At the micropyle end are three nuclei, the female gamete and two nuclei called the *synergidae*. At the opposite end are three *antipodal cells* and near the centre are two *polar nuclei*.

Fertilisation

When the pollen tube reaches the ovary it grows towards an ovule and enters it, usually through the micropyle. Having penetrated into the nucellus, possibly guided by the synergidae, the end of the pollen tube opens and the tube nucleus at its tip disintegrates. The two male gametes pass into the embryo sac. One fuses with the oosphere forming a diploid zygote; the second fuses with the two polar nuclei to form a *triploid endosperm nucleus*. The three antipodal cells take no part in fertilisation and often disappear soon after it has occurred.

Fruit formation

After fertilisation the ovules develop into seeds. Each zygote becomes an embryo consisting of a *radicle* (first root), *plumule* (first shoot), and one or two *cotyledons* (seed leaves). The triploid endosperm nucleus gives rise by a series of mitotic divisions to a mass of food storage tissue called *endosperm*. This is used by the embryo in its development and in an endospermic seed some remains in the seed as a food store. In non-endospermic seeds the cotyledons are the chief food store. Around the outside of the seed the integuments become the tough resistant seed coat or *testa*.

Finally the seed loses much of its water content, leaving it as low as 10 to 15% by weight of the seed contents, and in this desiccated state it remains dormant until the correct conditions are supplied for germination to take place.

While the seeds are developing the gynaecium begins to form a fruit. If other parts of the flower, such as the receptacle are also involved in the structure then it is termed a *false fruit* or *pseudocarp*, e.g. strawberry (Figure 87).

Fruit structure varies according to the form of the gynaecium and whether the ovary wall becomes dry and tough or soft and fleshy. There is usually some adaptation of the fruit for a particular method of dispersal, which reduces competition with the parent plant.

THE VARIETY OF FLORAL STRUCTURE

Reference has already been made to the great variety of structure and arrangement found in angiosperm flowers. The following pages give just a few examples. The investigation of the structure of a flower is completed by drawing a half-flower, floral diagram and floral formula. The half-flower represents a view of the flower cut along the median plane (in line with the main stem). All cut surfaces are shown by a double line. The floral diagram records the plan of the flower as if cut transversely. The bract ◡ is shown at the base of the diagram;

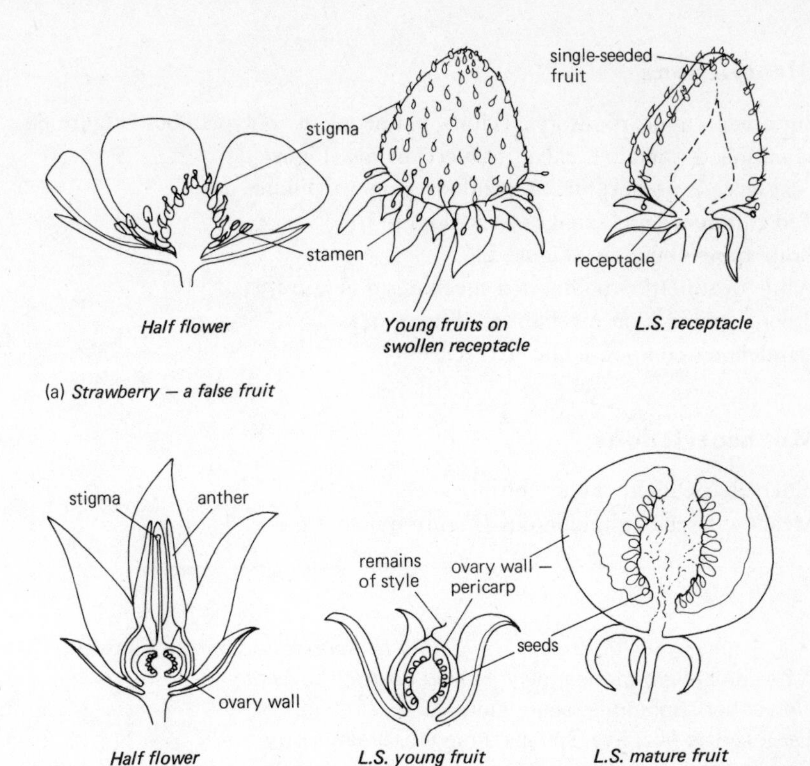

Figure 87 Stages in fruit formation

the peduncle by ○ at the top. Joined structures, such as petals or sepals are shown linked together. Stamens attached to petals are shown by linking radial lines. It is necessary to cut a transverse section of the ovary to see the number of loculi and placentation.

The floral formula is a coded description of the floral structure: K – calyx, C – corolla, A – androecium, G – gynaecium, e.g. K5 represents 5 sepals;

⊕ actinomorphic; ·|· or ↑ zygomorphic.

A superior ovary is shown by a line drawn below the number of carpels $G\underline{5}$ and an inferior ovary by a line above it $G\overline{5}$.
Parts joined are indicated by brackets K(3) + 2 (3 fused sepals; 2 free).
Linking lines show where one whorl is attached to another:

$\overset{\frown}{C6 \quad A6}$ (six stamens attached to six petals).

If perianth segments are present instead of sepals and petals, the letter P is used.

A list is given overleaf of the flowers that have been included and some of the features that make them particularly interesting:

Dicotyledons

Buttercup – apocarpous gynaecium, polysepalous, polypetalous (Figure 88).
Monkshood – petaloid calyx, reduced petals (Figure 89).
Foxglove – zygomorphic, epipetalous stamens (Figure 90).
Red campion – unisexual flowers (Figure 91).
Crab apple – epigyny (Figure 92).
White deadnettle – pollination mechanism (Figure 93).
Broom – pollination mechanism (Figure 94).
Dandelion – composite flower (Figure 95).

Monocotyledons

Bluebell – petaloid (Figure 96).
Meadow fescue – glumiflerous (Figure 97).

Figure 88 (opposite, above) Structure of *Ranunculus acris*, – Buttercup
A perennial herb of meadows. Flowering April to September.
Flower hermaphrodite, actinomorphic.
Calyx sepals five, free, spirally arranged, green.
Corolla petals five, free, golden yellow, glossy above, paler on lower surface.
Androecium forty to seventy stamens, spirally arranged, anthers dehisce towards the outside.
Gynaecium superior, apocarpous, about thirty free carpels spirally arranged on top of receptacle. Each carpel contains one ovule with basal placentation. Short style with curved stigma arises from each carpel.
Nectary at the base of each petal, covered by a small scale, under which nectar collects.
Pollination by flies and short-tongued bees. Outer stamens dehisce before the inner ones and bend outwards as they mature. Insects brush against anthers as they pass to reach nectaries. Self-pollination may occur in some flowers when the inner stamens shed pollen on to the stigmas or if small insects crawl over the flower before it matures.
Fruit a head of achenes – single carpels enclosing single seeds, which dry out and drop when mature.

Figure 89 (opposite, below) Structure of *Aconitum anglicum* – Monkshood
A perennial herb of damp, shady places, often by streams. Very local in SW England and Wales. Also cultivated in gardens. Flowering May to June.
Flower hermaphrodite, zygomorphic.
Calyx sepals five, free, spirally arranged, dark purple (petaloid), posterior sepal enlarged as a hood.
Corolla petals usually eight, free, two posterior petals elongated into stalks ending in spurs inside hood. Remaining petals, usually six, are very small and coloured purple.
Androecium about sixty stamens, spirally arranged, filaments blue, anthers greenish-blue, dehiscing towards the outside.

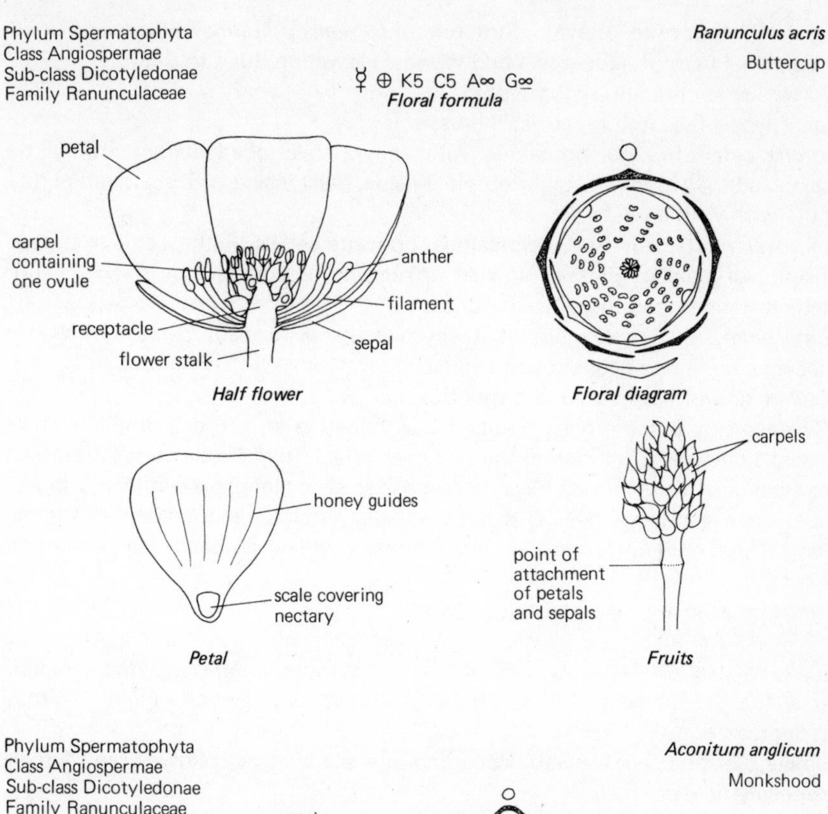

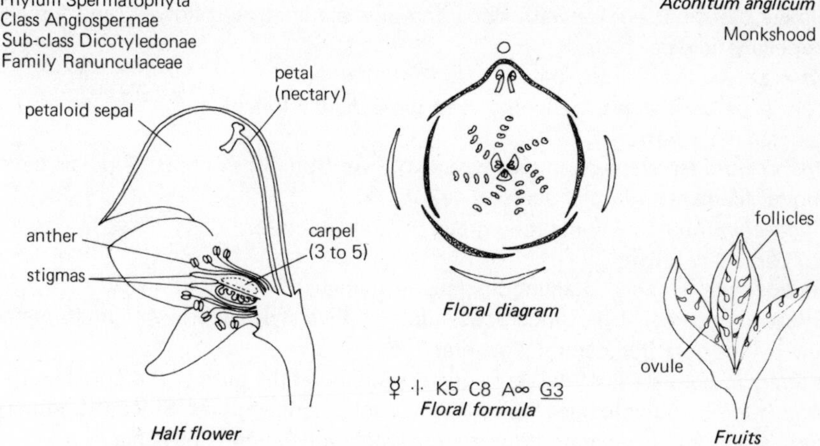

Gynaecium superior, apocarpous, three to five free carpels. Ovaries long, unilocular, ovules in two rows on inner side.

Nectary nectar secreted by spurs of posterior petals inside hood.

Pollination by long-tongued bumble bees. Flowers protandrous, immature stamens are curved back and curve forward towards the entrance of the hood as they mature. After the anthers have withered the stigmas mature. A bumble bee landing on the front and side sepals touches either the anthers or stigmas as it reaches up into the hood for nectar. Self-pollination is very unlikely.

Fruit three to five follicles, derived from single carpels dehiscing along their inner edges.

Figure 90 (opposite, above) Structure of *Digitalis purpurea* – Foxglove
A perennial herb of hedgerows and woods. Flowering June to July.
Flower hermaphrodite, zygomorphic.
Calyx sepals five, free except at the base.
Corolla petals five gamopetalous, tubular with five-lobed margin, front lobe making a broad lower lip. Light purple outside, paler inside with scattered purple spots each with a white rim.
Androecium stamens four, epipetalous, opposite sepals. Fifth posterior stamen absent. All stamens pressed against upper side of corolla tube. Two anterior stamens longer than the posterior ones.
Gynaecium superior, syncarpous, ovary bilocular with many carpels with axile placentation, long style, bilobed stigma.
Nectary ring of tissue around base of ovary.
Pollination by bumble bees. Spots act as honey guides. Protandrous, anterior stamens dehiscing first before the posterior ones. Stigma lobes remain pressed together until after stamens have dehisced then style elongates a little and stigma lobes open out. Large bees entering corolla will touch either anthers or stigma. Self-pollination is unlikely except in old flowers, when pollen may fall on to a ripe stigma.
Fruit a capsule dehiscing by two valves.

Figure 91 (opposite, below) Structure of *Melandrium dioicum* – Red campion
An annual or biennial herb of moist, shady hedgerows and woods. Flowering May to September.
Flower dioecious (unisexual), occurring as male staminate flowers and female carpellary flowers.
Calyx sepals five, gamosepalous, usually tinted red.
Corolla petals five, polypetalous, deep pink, deeply forked.
(a) staminate flower
Androecium ten stamens in two whorls of five. Outer ones opposite petals have shorter filaments. Introrse dehiscence.
Gynaecium non-functional, tiny green central structure.
(b) carpellary flower
Androecium ten small staminodes (sterile stamens).
Gynaecium carpels five, syncarpous, superior. Five styles and five stigmas, many tiny ovules on a free central placenta.
Nectary in both kinds of flower nectar is secreted at the base of the corolla tube.
Pollination by bumble bees, butterflies and long-tongued flies. Since the anthers and carpels are in separate flowers only cross-pollination is possible.
Fruit rounded capsule opening by ten curved 'teeth'.

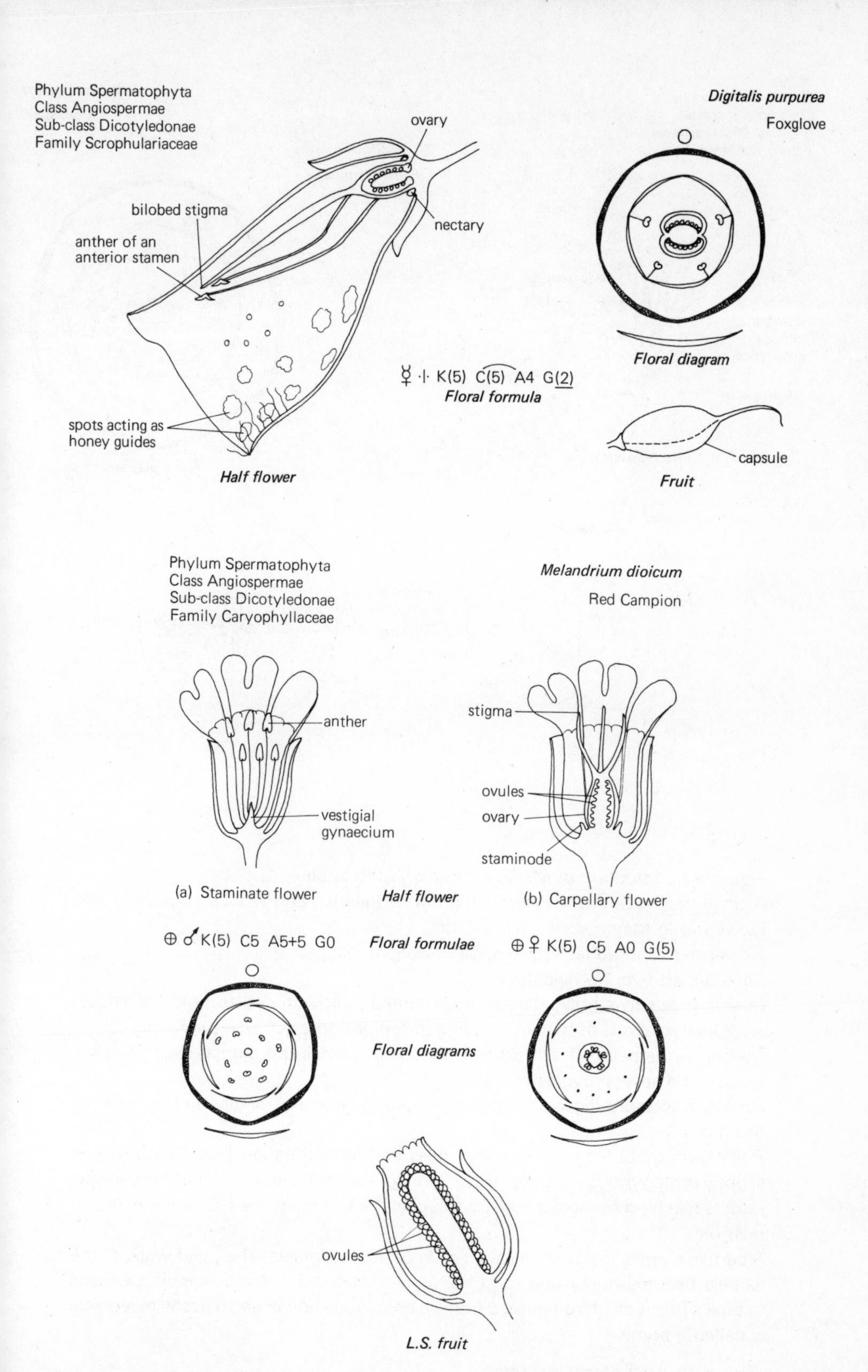
Phylum Spermatophyta
Class Angiospermae
Sub-class Dicotyledonae
Family Scrophulariaceae
Digitalis purpurea
Foxglove
ovary
bilobed stigma
nectary
anther of an
anterior stamen
spots acting as
honey guides
Half flower
☿ ·|· K(5) C(5) A4 G(2)
Floral formula
Floral diagram
capsule
Fruit
Phylum Spermatophyta
Class Angiospermae
Sub-class Dicotyledonae
Family Caryophyllaceae
Melandrium dioicum
Red Campion
anther
stigma
ovules
ovary
vestigial
gynaecium
staminode
(a) Staminate flower
Half flower
(b) Carpellary flower
⊕ ♂ K(5) C5 A5+5 G0
Floral formulae
⊕ ♀ K(5) C5 A0 G(5)
Floral diagrams
ovules
L.S. fruit

Phylum Spermatophyta
Class Angiospermae
Sub-class Dicotyledonae
Family Rosaceae

Malus sylvestris
Crab apple

☿ ⊕ K5 C5 A∞ $\overline{G(5)}$

Floral formula

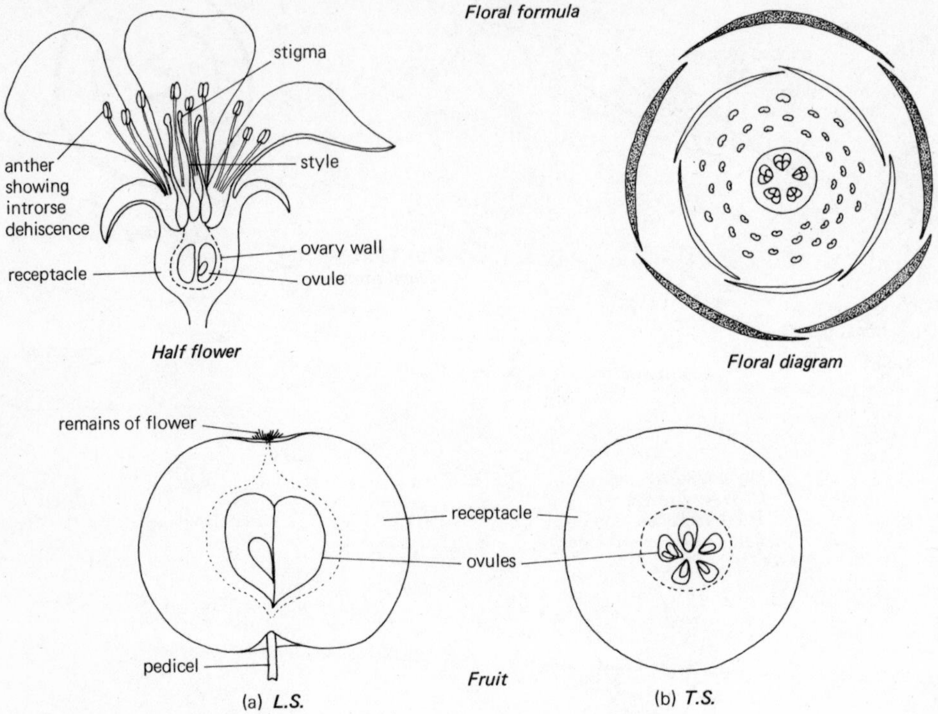

Half flower

Floral diagram

Fruit

(a) *L.S.*

(b) *T.S.*

Figure 92 Structure of *Malus sylvestris* – Crab apple

A small tree, seven to ten metres high in hedgerows and woods, especially oak. Flowering in May.

Flower hermaphrodite, actinomorphic, epigynous.

Calyx sepals five, polysepalous.

Corolla petals five, polypetalous, white tinted pink with darker pink undersides.

Androecium about forty stamens, free in four whorls, dehiscing inwards.

Gynaecium carpels five, syncarpous, inferior, five styles and stigmas. Ovary has five locules and several ovules in each.

Nectary Nectar produced on concave receptacle between stamens and the bases of the styles.

Pollination by bumble bees, honey bees and long-tongued flies. The flower is slightly protogynous – the stigmas becoming receptive before the anthers dehisce. Insects reaching for nectar may brush against the stigmas. Self-pollination is also possible.

Fruit five carpels inside the fleshy tissue of the receptacle. The inner walls of the carpels become horny and form the 'core', inside each are a number of seeds or 'pips'. This type of fruit formed from an epigynous flower and a fleshy receptacle is called a pome.

Phylum Spermatophyta
Class Angiospermae
Sub-class Dicotyledonae
Family Labiatae

Lamium album

White deadnettle

anther of posterior stamen

bilobed stigma

style

filament of anterior stamen

sepal

lower lip of corolla

ovary

fringe of hairs protecting nectary

Half flower

Floral diagram

☿ ·|· K(5) C(5) A4 $\underline{G(2)}$

Floral formula

four 1-seeded nutlets with hard walls

persistent calyx

Fruits

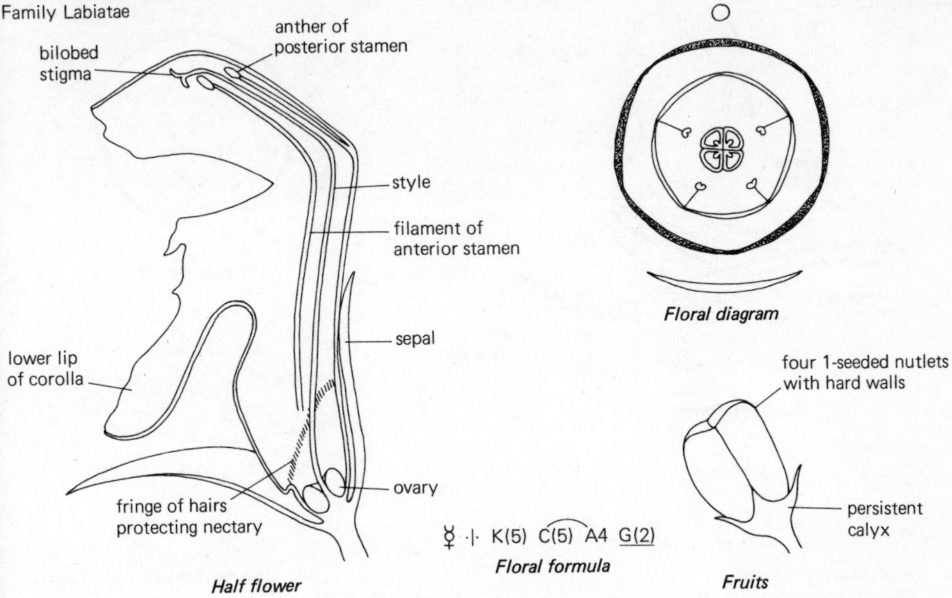

Figure 93 Structure of *Lamium album* – White deadnettle

A perennial herb of hedgerows. Flowering April to June and autumn.

Flower hermaphrodite, zygomorphic.

Calyx sepals five fused (gamosepalous) to form a narrow tube ending in five points.

Corolla petals five fused (gamopetalous), two-lipped, one large anterior petal forming lower lip, two lateral petals and two posterior petals fused as the hooded upper lip.

Androecium stamens four, epipetalous, opposite sepals, fifth posterior stamen absent. Anterior pair of stamens have longer filaments than the posterior pair.

Gynaecium superior, syncarpous, two carpels. Long style bearing two-lipped stigma lying between the anthers. Ovary initially bilocular, becoming quadrilocular by development of a false septum. One ovule in each loculus.

Nectary two swellings on anterior side of the ovary base.

Pollination by bumble bees. A bee alighting on the lower lip of the corolla fills the space between the two lips of the corolla. It touches the sigma first and may rub pollen on to it, then it touches the anthers. Self-pollination is also possible since anthers and stigma ripen at the same time.

Fruit four nutlets – two from each carpel, enclosed by a persistent calyx.

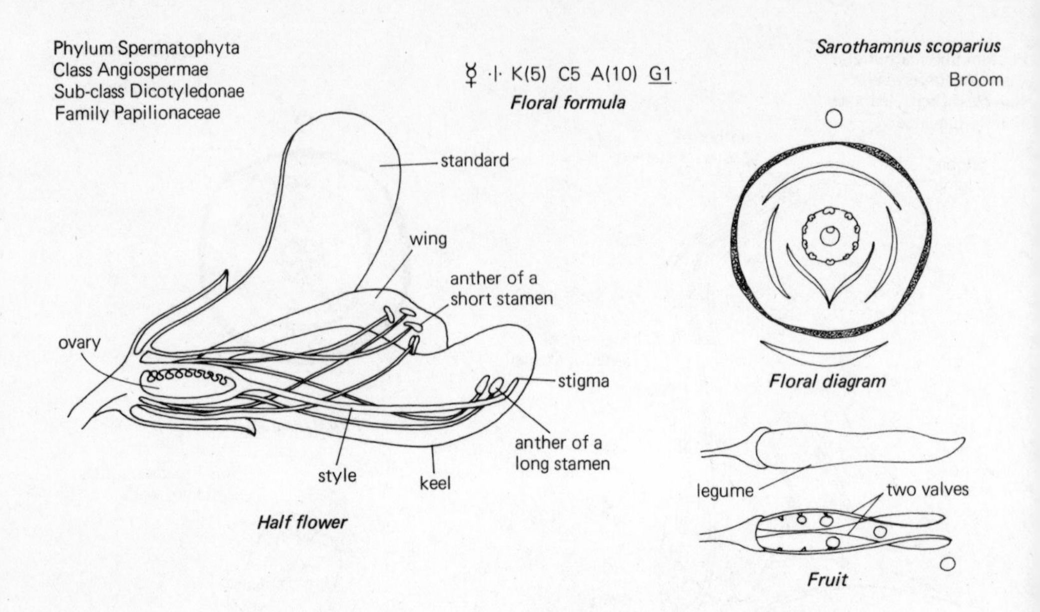
Phylum Spermatophyta
Class Angiospermae
Sub-class Dicotyledonae
Family Papilionaceae
☿ ·|· K(5) C5 A(10) G1
Floral formula
Sarothamnus scoparius
Broom
standard
wing
anther of a short stamen
ovary
stigma
anther of a long stamen
style
keel
Half flower
Floral diagram
legume
two valves
Fruit

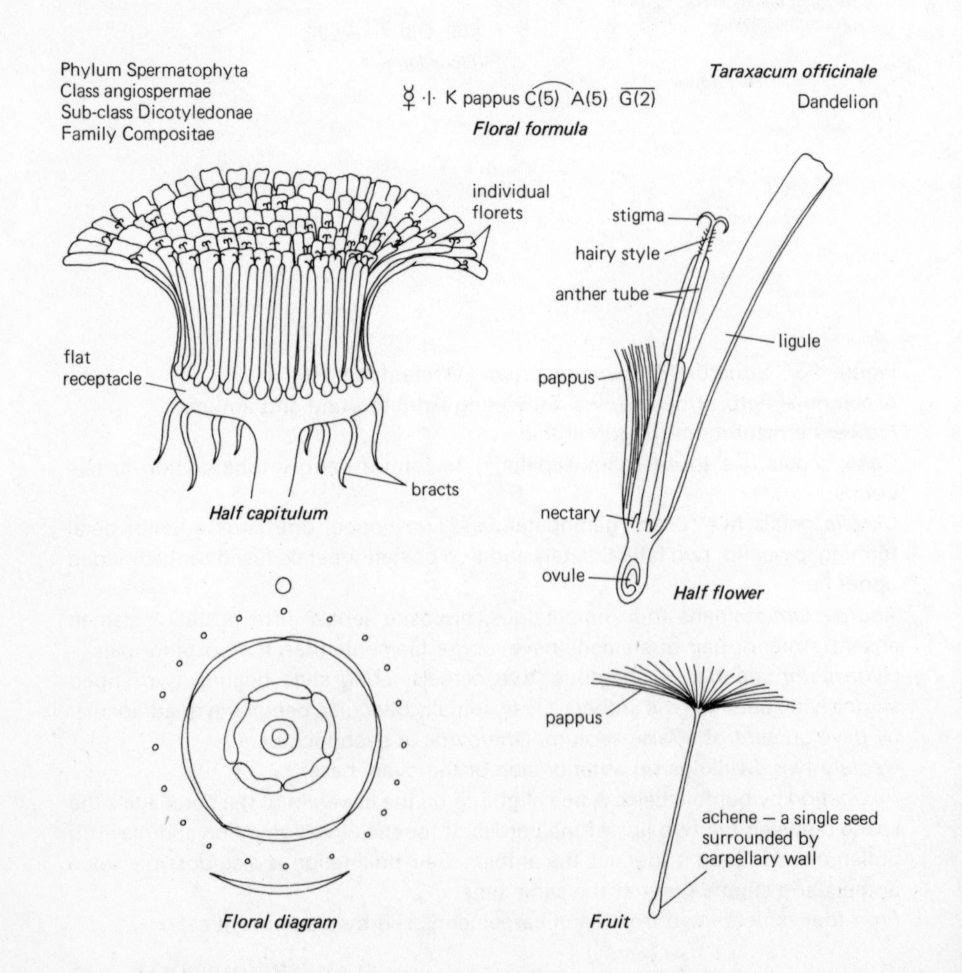
Phylum Spermatophyta
Class angiospermae
Sub-class Dicotyledonae
Family Compositae
☿ ·|· K pappus C(5) A(5) G(2)
Floral formula
Taraxacum officinale
Dandelion
individual florets
flat receptacle
bracts
Half capitulum
stigma
hairy style
anther tube
ligule
pappus
nectary
ovule
Half flower
pappus
achene – a single seed surrounded by carpellary wall
Floral diagram
Fruit

Figure 94 (opposite, above) Structure of *Sarothamnus scoparius* – Broom
A shrub, 1 to 2 m high, on warm, dry soil of heaths and commons. Flowering May to June.
Flower hermaphrodite, zygomorphic.
Calyx sepals five, gamosepalous, two-lipped, upper lip with three teeth, lower lip with two teeth.
Corolla petals five, polypetalous, yellow, a posterior standard, two lateral wing petals, two anterior petals fused as a keel.
Androecium ten stamens, filaments all fused in to a tube around ovary (monadelphous). Five short stamens and five longer ones.
Gynaecium single carpel, superior, unilocular, several ovules on a marginal placenta.
Nectary none. These are pollen flowers.
Pollination by honey bees and bumble bees. Short stamens mature first and drop pollen into the keel. The weight of a bee landing on the lateral petals depresses them and the keel. The keel petals separate allowing the stamens and style to suddenly spring upwards, throwing pollen over the bee. The short stamens hit the underside of the bee and the long stamens and style hit its back. Self-pollination seems quite likely but it seems that most brooms are self-sterile.
Fruit a legume or pod, dehiscing violently into two valves and twisting, flinging out the seeds.

Figure 95 (opposite, below) Structure of *Taraxacum officinale* – Dandelion
A perennial herb of grassland and waste places, flowering throughout the year but mainly March to October.
Flower hermaphrodite, zygomorphic, one to two hundred flowers or florets on a flat receptacle forming an inflorescence called a capitulum.
Calyx a pappus of many soft white hairs.
Corolla petals five, united into a tube at the base and extending as a bright yellow strap or ligule (ligulate).
Androecium five stamens, epipetalous, anthers joined as a tube (syngenesious).
Gynaecium inferior, syncarpous, two carpels but only one persisting. Ovary unilocular with one basal ovule, long style, bilobed stigma.
Nectary nectar secreted around base of style.
Pollination by all types of bees, flies and butterflies since nectar is easily accessible. Florets mature from the outside inwards. They are protandrous, pollen being released into anther tube before the florets are even open. The style is short and the stigma lobes are pressed together. Later it elongates and pushes pollen up the anther tube. The stigma lobes gradually open out and curl round to touch the anthers, thus allowing self-pollination if cross-pollination has failed.
Fruit a head of achenes each with a pappus 'parachute', dispersed by wind.

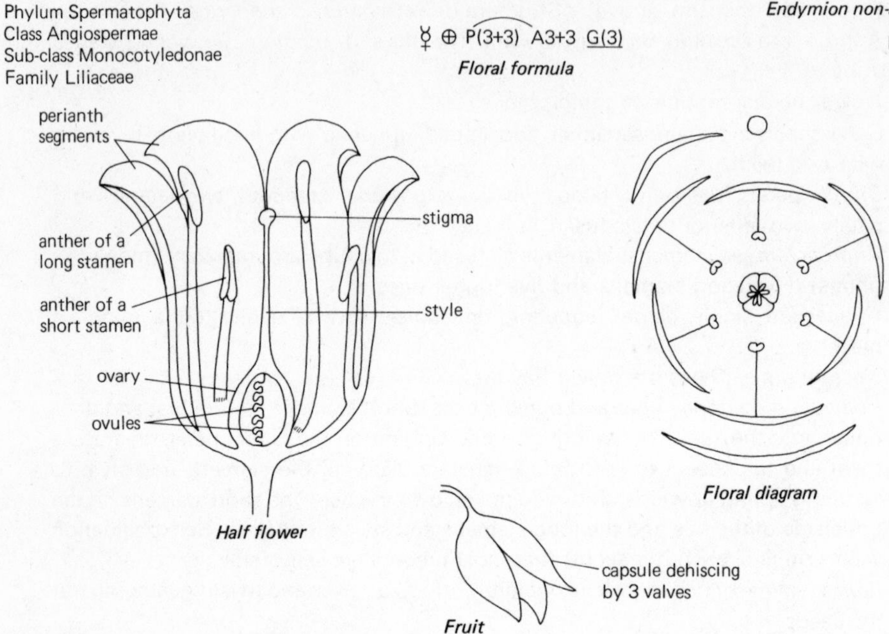

Figure 96 Structure of *Endymion non-scriptus* – Bluebell

A perennial herb with bulb. Woods and shady hedgerows. Flowering April to May.

Flower hermaphrodite, actinomorphic.

Perianth no distinction into sepals and petals. Six perianth segments, petaloid, deep purple, arranged in two whorls of three, free except right at the base, forming a bell shape. Slight fragrance.

Androecium six stamens borne on the perianth segments. The outer stamens have longer filaments and are attached higher up the perianth segments; inner stamens have shorter filaments and are attached nearer the base of the flower.

Gynaecium superior, syncarpous, three carpels, trilocular, many ovules on axile placentae. Long style ending in a three-lobed stigma.

Nectary nectar is produced near the top of the ovary.

Pollination by honey bees. The flower is slightly protogynous. Pollen may be rubbed on to a receptive stigma by a bee probing for nectar. Later, as the anthers mature, the style elongates and carries the stigma above them. While it is close to the anthers self-pollination is possible.

Fruit a capsule dehiscing into three valves, releasing shiny black seeds.

Figure 97 (opposite) Structure of *Festuca pratensis* – Meadow fescue

A perennial grass 30 to 120 cm tall, often in tussocks. Found in meadows, low-lying grassland and on roadside verges.

Flowers hermaphrodite, zygomorphic, with glumes.

Glumes persistent, lance-shaped, with membranous tips and edges.

Lemma lance-shaped, without awns (stiff, hair-like projections found in some grasses).

Phylum Spermatophyta
Class Angiospermae
Sub-class Monocotyledonae
Family Gramineae

Festuca pratensis

Meadow fescue

·|· P0 A3 G(3)

Floral formula

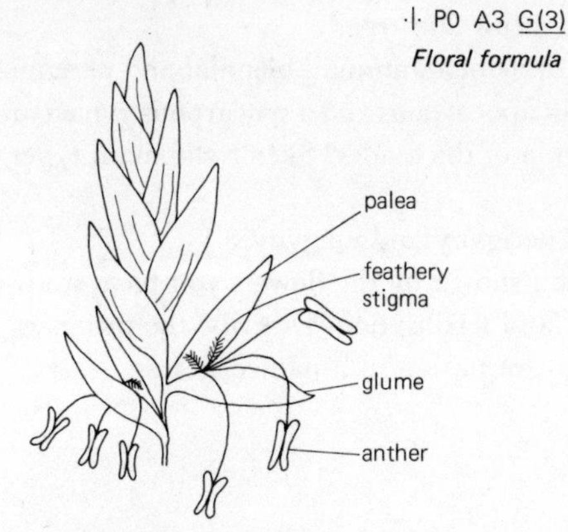

A spikelet with two mature flowers

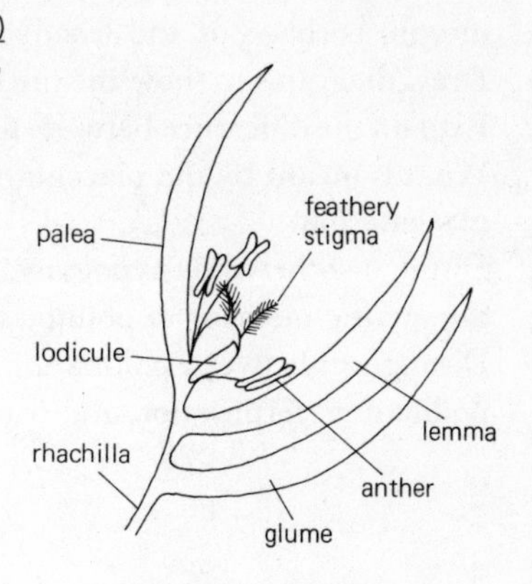

Diagram to show the relative position of the floral parts

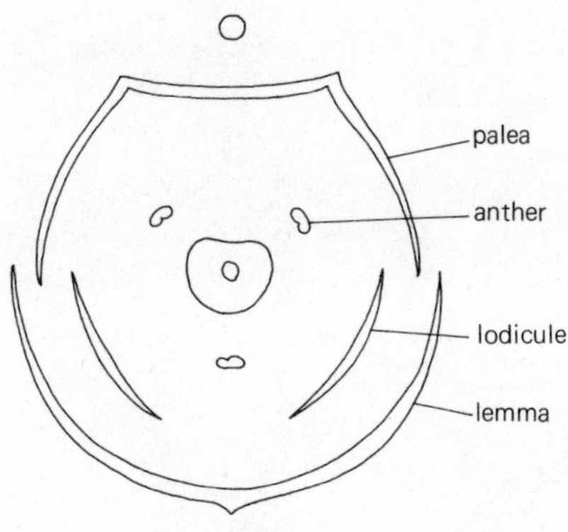

Floral diagram

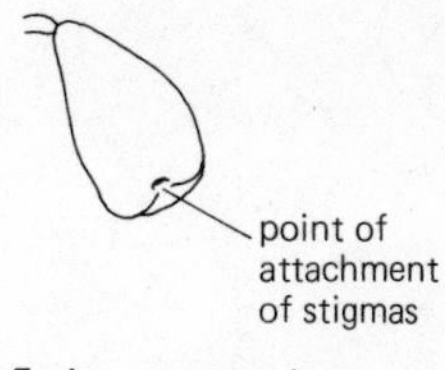

Fruit – a caryopsis

Palea same length as the lemma, keeled.
Lodicules two, rounded at the base and having two points at the top. Arranged one on each side of the ovary. They may be the remains of a perianth, the rest of which has been suppressed.
Gynaecium superior, three carpels, only one remaining functional. Syncarpous with a single locule containing a single ovule. Two short styles bearing long feathery stigmas.
Nectary none.
Pollination by wind. The anthers hang outside the paleae where they are easily shaken by the wind. The feathery stigmas are also well-exposed and receive wind-blown pollen.
Fruit a single-seeded dry fruit in which the ovary wall (pericarp), becomes fused to the testa of the seed forming a grain or caryopsis.

QUESTIONS

1 Distinguish between: hydrophyte and mesophyte; xerophyte and xeromorph; herbaceous and woody; shrub and tree.
2 Draw diagrams to show the life cycle of an annual, biennial and perennial.
3 Explain the difference between an apocarpous and a syncarpous gynaecium. What is meant by the placentation of the ovules? Sketch the main types of placentation.
4 Distinguish between hypogyny, perigyny and epigyny.
5 Survey the methods of pollination shown by the flowers you have studied.
6 Distinguish between pollination and fertilisation. Describe the processes of pollination, fertilisation and fruit formation in a named flower.

18 Phylum Chordata

Chordates first appeared in the Ordovician period some 450 million years ago. This phylum includes the familiar vertebrate classes and several less well-known invertebrate classes.

Chordates generally exhibit a greater complexity of structure and a higher degree of organisation than any other group in the animal kingdom. This increased complexity has necessitated an increase in size, and the mean size of chordates is accordingly larger than the mean size in any other phylum.

Chordates are found on land, in the sea and freshwater, and also in the air. They are sometimes described as the dominant form of animal life, but it should be realised that this refers more to their size and complexity and their evolutionary position than to numbers of species and numbers of individuals.

GENERAL FEATURES

Chordates are bilaterally symmetrical, triploblastic, metamerically segmented coelomates. In addition most chordates possess the following features for at least part of their life history (Figure 98):

1 a *notochord* – this is a dorsally positioned, rod-like structure made of a substance similar to cartilage. It is present in the embryos of all chordates where it provides support. In vertebrate classes the notochord is replaced by the vertebral column during embryonic development;

2 a tubular, dorsal nerve cord from which there normally arise segmentally arranged, paired, peripheral nerves that pass to all parts of the body;

3 the pharynx is perforated by paired slits, the *pharyngeal clefts*, which pass to the outside of the body. During the course of evolution the function of the pharyngeal clefts has changed. In the invertebrate sea-squirts and lancelets the pharyngeal clefts form the filter-feeding apparatus; in fish they develop into vascularised gill slits and serve a respiratory function; while in terrestrial chordates the pharyngeal slits present in the embryo disappear during the course of development and only the first pair persist in the adult as the eustachian tubes connecting the middle ear with the back of the throat;

4 segmentally arranged muscle blocks or *myotomes* are present in the embryonic stages of most chordates, but are distinguishable in the adults of only the more primitive chordates, namely lancelets and fish;

5 a closed blood vascular system with blood flowing forwards to the heart in the main ventral vessel and away from the heart in the main dorsal vessel. This contrasts with the situation in non-chordates where the reverse applies;

6 a segmented tail positioned behind the anus.

The chordates are believed to have evolved from an echinoderm larva by *paedogenesis* – a process wherein one species gives rise to another by developing the ability to reproduce whilst still a larva.

CLASSIFICATION

The chordates are classified as follows:

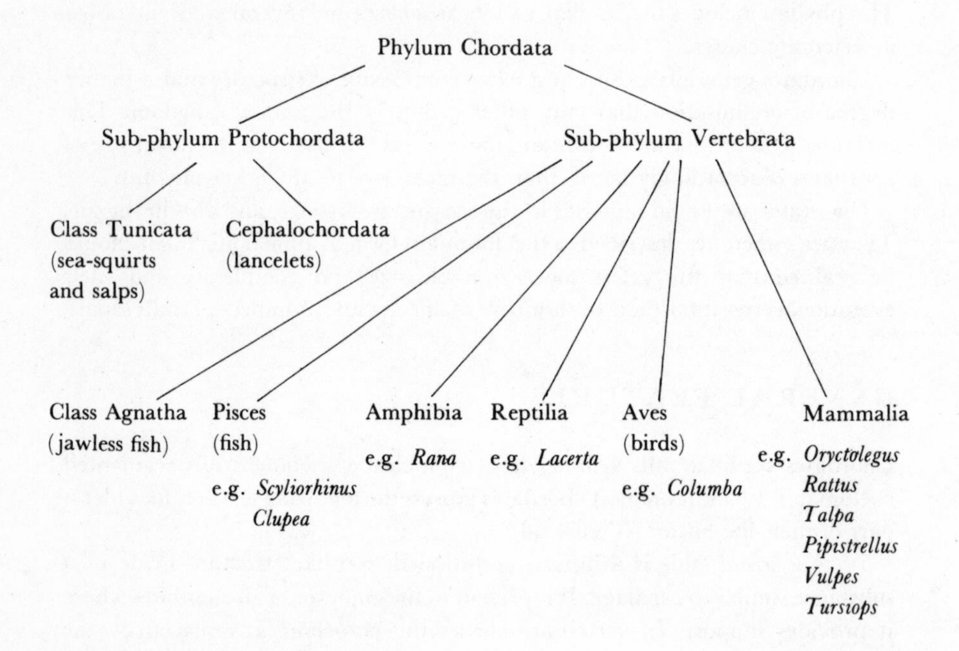

Sub-phylum Protochordata

This group includes all the non-vertebrate (invertebrate) members of the phylum. All species are marine, filter-feeding animals. Sea-squirts are mainly sedentary and often colonial, while salps are mostly colonial, planktonic forms. Lancelets are small fish-like animals, most of which inhabit shallow coastal waters.

Sub-phylum Vertebrata

This group comprises the majority of chordates. They have many features in common. In all vertebrates the notochord is replaced, at least in part, by a vertebral column or backbone, made of cartilage or bone, which surrounds the nerve cord. Associated with the backbone there are very often two girdles, an anterior pectoral girdle and a posterior pelvic girdle; these serve for the attachment of the fins or limbs.

In all vertebrates, except the agnathans, jaws develop around the mouth. The

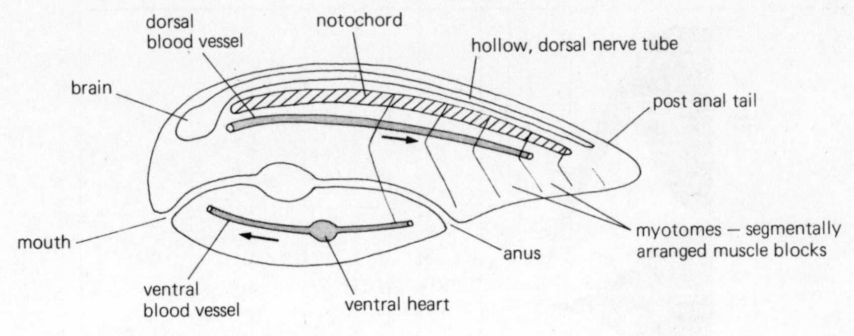

Arrows indicate direction of blood flow

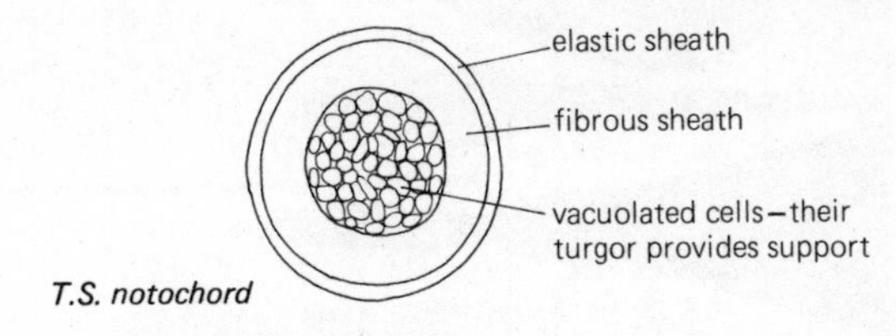

Figure 98 Diagram of a generalised chordate

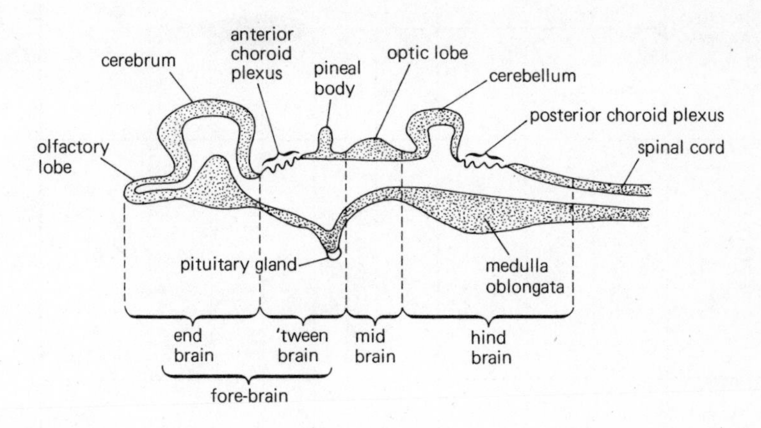

Figure 99 Diagrammatic longitudinal section through a dogfish brain showing the main regions and the structures associated with each region. (Redrawn after Roberts, *Biology – a functional approach*, Nelson, 1971.)

---

brain is surrounded by a bony or cartilaginous case, the cranium. The brain itself has the same basic pattern in all vertebrates; differing mainly in the size and complexity of the various regions. This basic vertebrate plan is seen most clearly in the fish brain (Figure 99). Sense organs – paired eyes, ears and olfactory organs – develop as outgrowths of the brain or in association with it.

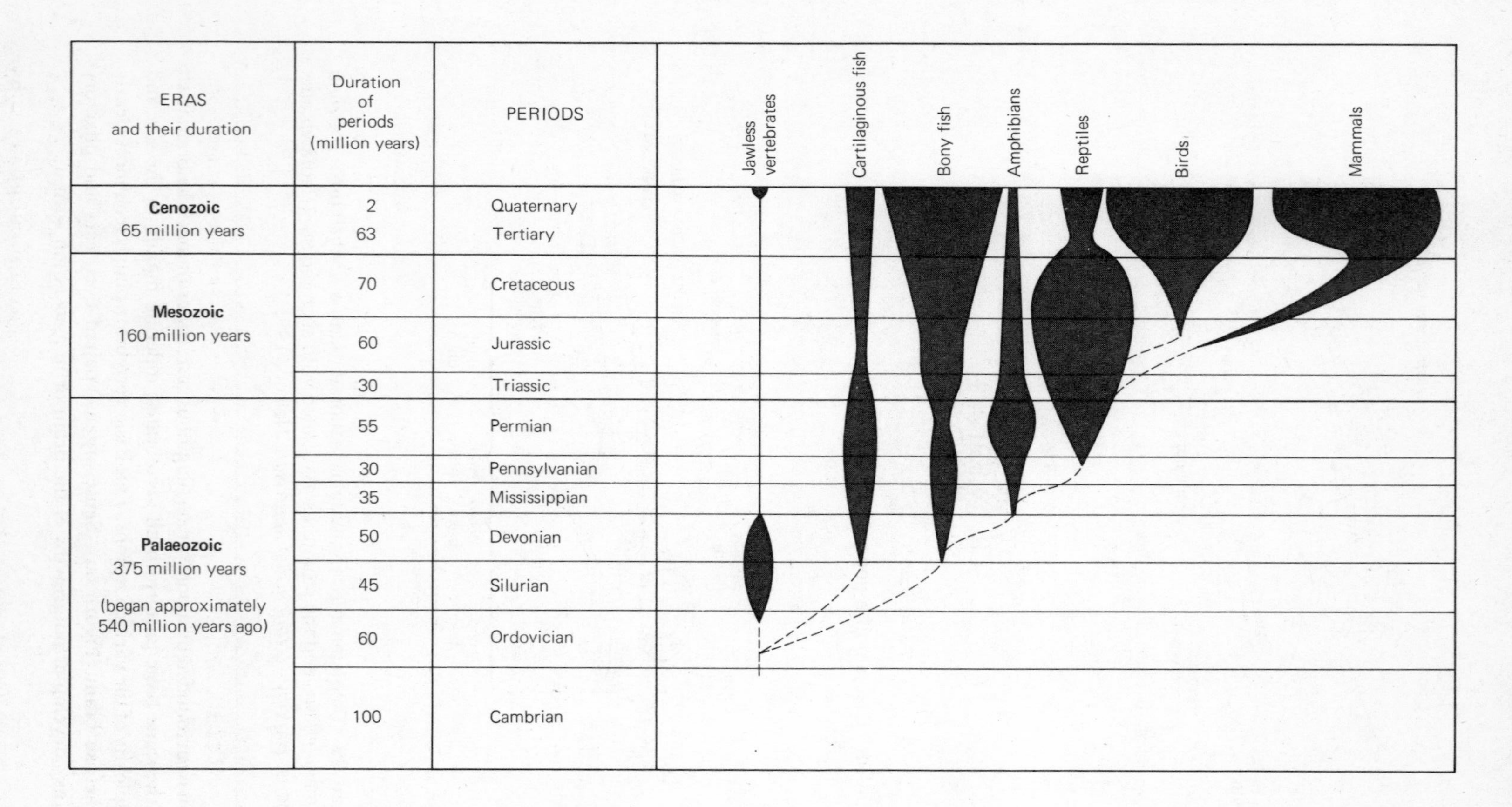

Figure 100 Geological time chart showing the emergence of the main vertebrate classes and their relative abundance through time

Vertebrates all have a well developed, closed, blood vascular system. Circulation of the blood is maintained by a muscular chambered heart. The blood contains corpuscles, some of which incorporate the red, oxygen-carrying pigment, haemoglobin. This contrasts with the situation in invertebrates, where haemoglobin occurs only infrequently, and then it is usually in solution in the plasma, rather than in special cells.

All vertebrates possess two kidneys, excretory organs composed of numerous tubules or *nephrons*.

Fossils of vertebrates are relatively abundant. This is partly attributable to their hard skeleton which fossilises comparatively easily. These fossils have enabled us to build up a fairly complete picture of the evolution of this group (Figure 100).

The main classes of vertebrates will now be described.

Class Pisces

The fossil record shows that fish are the most primitive vertebrates. They first appeared in the Ordovician and by the middle of the Devonian period, a little over 50 million years later, so many species had evolved that the Devonian became known as the 'age of fishes'.

All fish are aquatic, although a few species can spend limited periods out of the water. In most fish the body is streamlined, with smooth contours and few projecting organs, adapted for moving through water with the minimum resistance. All fish possess fins, these consist of a strong membrane supported by bony or cartilaginous rays. The fins confer stability and the ability to manoeuvre, and in a few species they provide the propulsive thrust in locomotion. In most fish movement is brought about by waves of contraction that pass along the segmentally arranged muscle blocks on either side of the backbone.

Fish are *poikilotherms* – their body temperature varies according to the temperature of the surroundings. The organs of gas exchange are the gills. Each gill is composed of a series of leaf-like filaments, the *primary lamellae*, which bear folds, the *secondary lamellae*. The surface for gas exchange is thus very extensive.

Circulation in fish is single – the blood passes through the heart only once in a complete circuit of the body. The heart has two chambers, an atrium and a ventricle (Figure 101).

The sexes in fish are separate and fertilisation is generally external. Few fish exhibit parental care and reproduction is usually prolific to compensate for the high mortality of young fish.

The class Pisces is divided into two sub-classes, the cartilaginous fish or Chondrichthyes, and the bony fish or Osteichthyes. As their names suggest, the main feature distinguishing the two is the composition of the skeleton. The main differences between the two sub-classes are set out in Table 3.

Sub-class Chondrichthyes

There are about 3000 present day species in this group and many more extinct

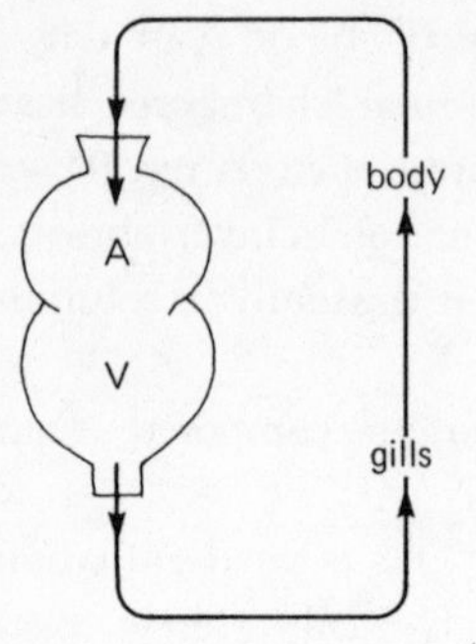

Fish – single circulation with undivided heart

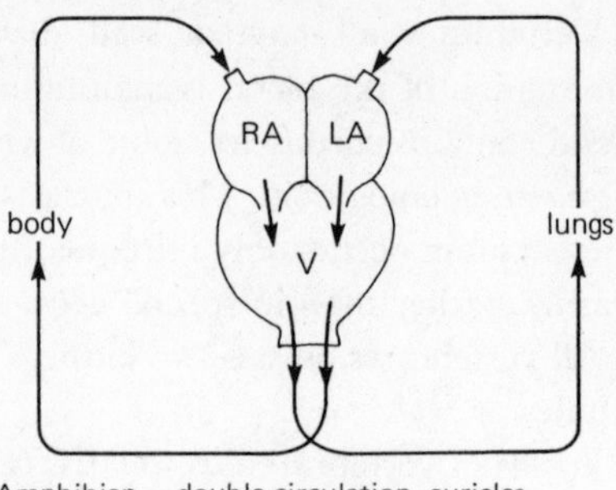

Amphibian – double circulation, auricles divided by septum, ventricles undivided

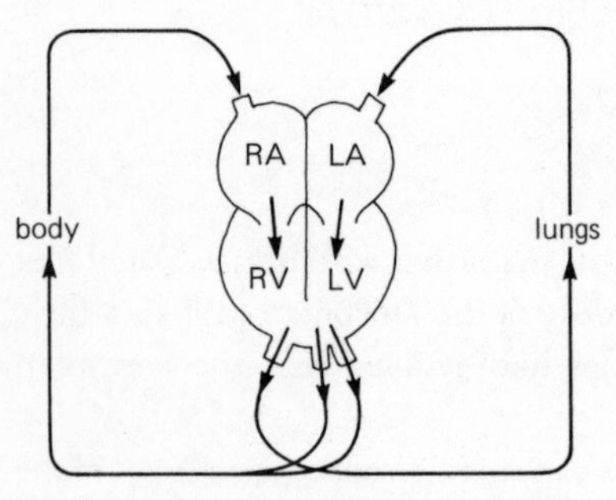

Reptile – double circulation auricles divided ventricles partially divided by septum

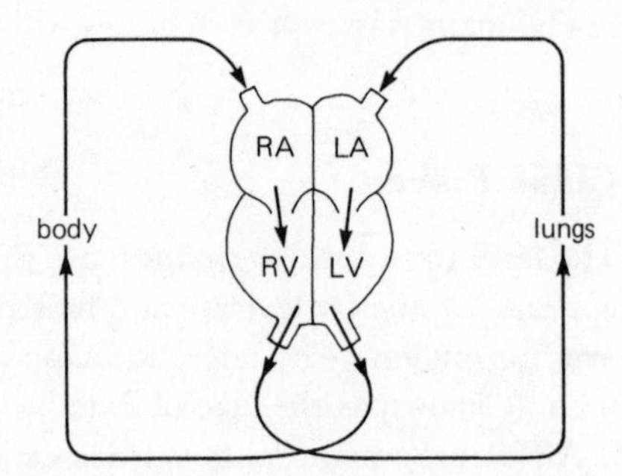

Birds and mammals – double circulation with completely divided heart

Figure 101 Plans of the heart and circulation in different vertebrates
Although the ventricles are not apparently separated in amphibians and reptiles, experiments have shown that little if any mixing of oxygenated and deoxygenated blood occurs in the ventricle. (A = auricle, V = ventricle, R = right, L = left, → indicates direction of blood flow)

species. Extant (living) forms include the familiar rays, sharks, skates and dogfish. Nearly all members of this group are highly specialised marine predators. Exceptions include the whale shark and the manta ray, both very large plankton-feeders. The whale shark is in fact the largest living fish, reportedly having a maximum length of around 20 m. and a weight of 20 metric tons or more.

The common dogfish, *Scyliorhinus caniculus* is illustrated in Figure 102.

Sub-class Osteichthyes

Many different types of bony fish evolved during the Mesozoic. Towards the end of that era, one group, the teleosts, began to diversify and rapidly expanded to become the dominant form of bony fish. Such an expansion, whereby one group of organisms develop specialisations which enable it to exploit a wide variety of habitats, is known as *adaptive radiation*.

Table 3 The main differences between the sub-classes *Chondrichthyes* and *Osteichthyes*

	Chondrichthyes	*Osteichthyes*
Skeleton	composed entirely of cartilage	composed of bone and cartilage
Paired fins	very limited range of movement, cannot be rotated to act as brakes	more mobile, can usually be rotated to act as brakes
Tail	dorsal lobe of tail fin usually much larger than vertral lobe, such a tail is termed *heterocercal*	dorsal and ventral lobes usually of equal size – *homocercal*
Skin	contains dermal denticles, these have a structure similar to that of teeth having a central pulp cavity surrounded by dentine with an outer covering of enamel	contains scales – bony plates covered by skin
Gill slits	open directly to outside	covered by a bony flap, the operculum
Position of mouth	ventral	usually dorsal
Swimbladder – gas-filled sac that lies under backbone	absent, must keep swimming to avoid sinking	present – functions as hydrostatic organ, enables fish to adjust its density so allowing it to remain at a certain depth
Intestine	spiral valve to increase surface area	spiral valve not normally present
Habitat	entirely marine	marine and freshwater

The bony fish are the most varied of all the vertebrate classes. There are more than 20000 species ranging in size from the tiny pygmy goby, which is only about 1 cm long, to the huge Russian sturgeon, which can reach 8.5 m in length. They exhibit tremendous diversity of form. The ribbon fish with its long flat body, the globe-shaped parrot fish with its bird-like beak, and the bizarre sea dragon which, with its leaf-like appendages, looks more like a piece of seaweed than a fish, are just a few of the many strange forms encountered in this group.

Unlike cartilaginous fish, bony fish abound in both marine and freshwater habitats. Within these habitats they have adapted to a wide range of conditions. The frozen streams of sub-arctic regions, the hand-hot springs of Arabia, and the ocean floor 12000m down, where the pressure is around 1.1×10^8 Pa (1100 atmospheres), are all inhabited by species of bony fish.

The herring, *Clupea harengus*, is illustrated in Figure 103.

Phylum Chordata
Sub-phylum Vertebrata
Class Pisces
Sub-class Chondrichthyes

Scyliorhinus caniculus

Lesser spotted dogfish

(a) *Adult – side view*

Length up to 70 cm

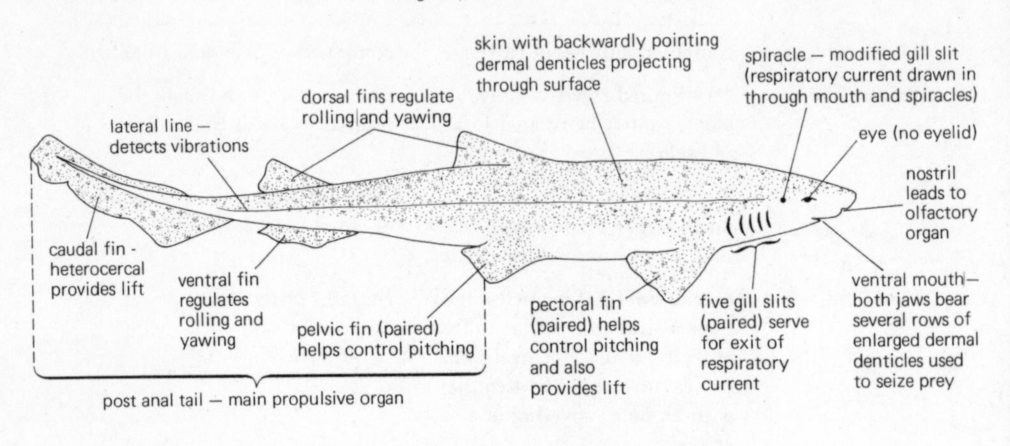

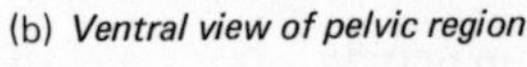
(b) *Ventral view of pelvic region*

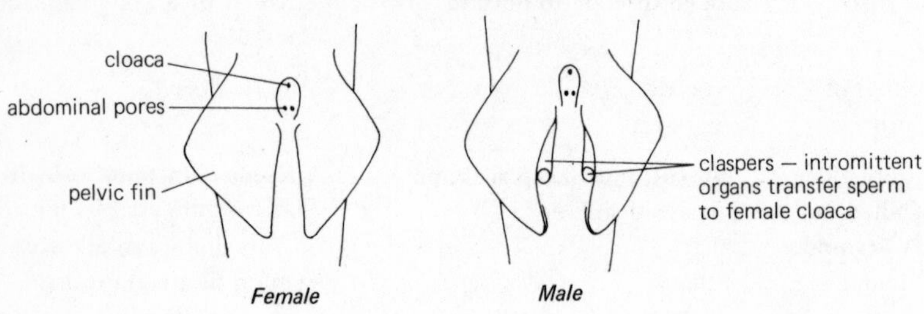

Figure 102 *Scyliorhinus caniculus*, the common or lesser spotted dogfish is a bottom living species that is very abundant in British coastal waters. Large numbers are caught and sold as rock salmon.

They feed on a variety of living and recently dead organisms, especially Crustacea and small fish. Dogfish are unusual among fish in that mating occurs. The male winds around the female and transfers sperm by inserting the claspers one at a time into the cloaca of the female. The eggs, known as mermaid's purses, are laid in flat, oblong cases with long threads at each corner. The threads serve to attach the eggs to a seaweed. Incubation lasts about six months after which the young fish emerge.

Class Amphibia

Amphibians first appeared at the end of the Devonian period. They are believed to have evolved from a group of fish with peculiar lobed or limb-like fins, known as crossopterygians. This group was thought to be extinct until 1938 when a specimen of *Latimeria*, the coelacanth, was caught off the coast of South Africa.

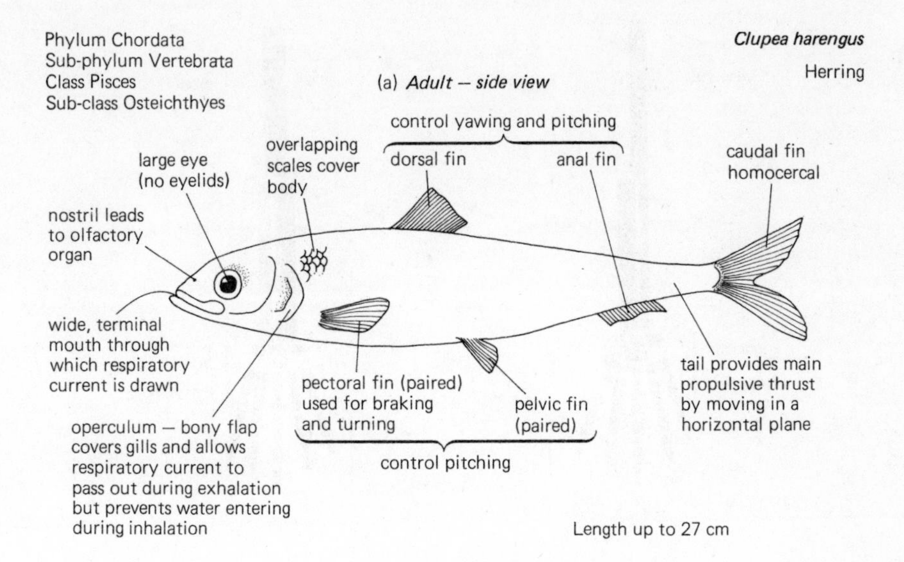

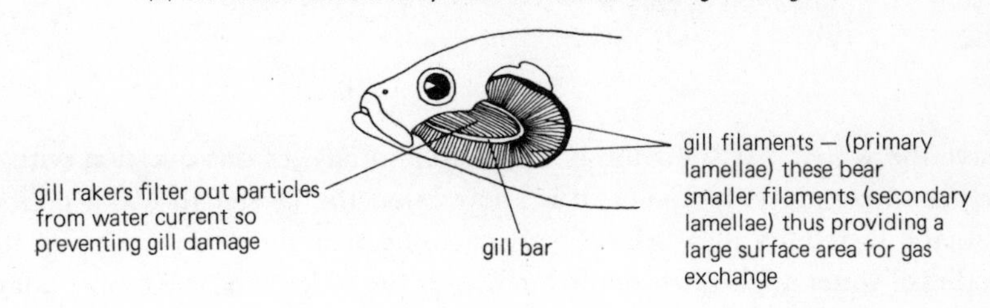

Figure 103 *Clupea harengus*, the herring, feeds, migrates and spawns in large shoals. They are carnivores and catch their prey selectively, prefering the crustacean *Calanus*, the arrow worm *Sagitta*, and sand eels.

Herrings undergo vertical migration. During the day they tend to remain in deep waters and at night they move up towards the surface.

Spawning occurs in autumn. The females shed ribbons of eggs onto the sea bed and the males then release sperm over the eggs. A single female can lay up to 20000 eggs per year but few will develop to maturity. The larvae hatch out after about two weeks. Soon afterwards they swim up into the planktonic zone where they remain for three to six months while they develop into adults.

Herrings are of considerable economic importance. Large numbers are caught annually; most are either smoked and sold as kippers or preserved in brine.

The arrangement of the fin bones in crossopterygians is thought to represent an early stage in the evolution of the pentadactyl or five-digit limb common to all terrestrial vertebrates, (see Figure 104).

As yet we have not considered the question 'why did the amphibian ancestors

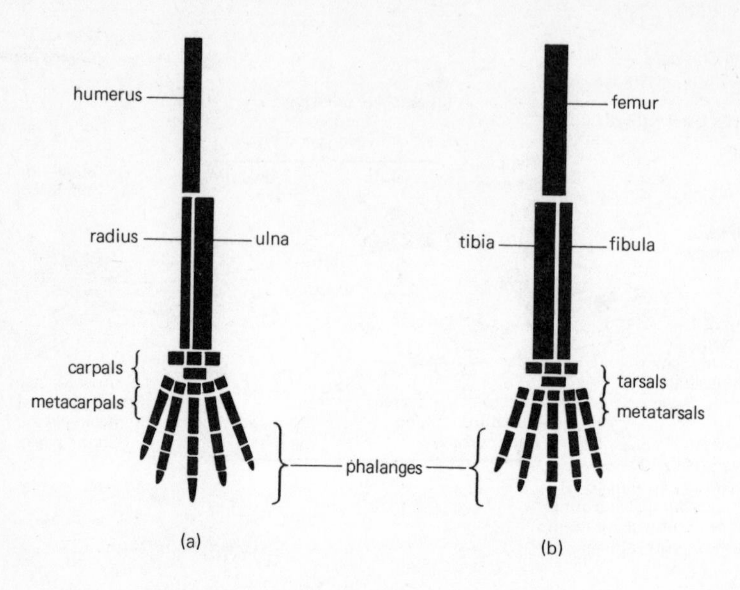

Figure 104 Diagram of the pentadactyl limb skeleton (a) fore-limb and (b) hind limb

leave the water?'. It is obviously impossible to answer this question with any degree of certainty. However, it is known that the Devonian was a period of climatic instability and it has been suggested that the periodic drying up of bodies of water might have tended to favour the survival of any species adapted to withstand exposure to air and perhaps able to move from one body of water to another.

The change from an aquatic to a terrestrial existence necessitated fairly extensive changes in body structure; the most important of these are outlined below.

Adaptations to life on land

1 The skin must be adapted to withstand drying. In Amphibia, as in all terrestrial vertebrates, this is accomplished by the deposition of the water-proofing, protein material, keratin, in the epidermis.

2 Gills are unsuitable for gas exchange in air as they are too delicate and would collapse and dry out. Thus in terrestrial vertebrates, gills are replaced by internally situated, sac-like, gas exchange organs, the lungs. These were probably already present in the crossopterygian ancestors of Amphibia, having developed from the swimbladder.

3 The skeleton has to take over the role of support which in aquatic organisms is largely provided by the water surrounding them. The skeleton is accordingly strengthened, particularly the backbone, girdles and limbs. Associated with this there is also the development of a more powerful and extensive musculature.

The earliest amphibians were rather large animals compared with modern forms; some are known to have reached lengths in excess of 6 metres.
Present day amphibians are divided into three orders:
ORDER APODA – legless, burrowing forms;
ORDER CAUDATA – tailed amphibians, newts and salamanders;
ORDER ANURA – tail-less forms, frogs and toads.
Amphibians, like fish, are poikilotherms. Their skin is usually fairly thin and moist, being well supplied with mucus glands.
They have a *double circulation*, that is the blood passes twice through the heart in each complete circuit of the body (Figure 101). The significance of this arrangement, which is found in all terrestrial vertebrates is that it improves the efficiency of the circulation. It means that blood, that has been slowed by passing through the capillaries of the lung, receives a 'boost' on returning to the heart, so that it passes to the rest of the body at a much faster rate and a much higher pressure than would otherwise be possible. The heart is three-chambered, with two atria and only one ventricle, although it appears that folds in the ventricle wall keep the oxygenated and deoxygenated blood separate.

With very few exceptions, amphibians return to the water to breed. This dependence on water has undoubtedly limited the expansion of this group. The eggs are laid and fertilised in the water. They hatch into aquatic larvae with external gills. These undergo a gradual *metamorphosis*, slowly changing into the adult which is typically terrestrial and breathes using lungs.

The common frog, *Rana temporaria*, is shown in Figure 105, and its life cycle is illustrated in Figure 106.

Class Reptilia

Reptiles are thought to have evolved from amphibians during the Carboniferous period. A group of fossil species, belonging to the genus *Seymouria*, provides evidence for this – they seem to have been an 'intermediate' type, possessing both reptilian and amphibian features.

The early reptiles were better adapted to terrestrial conditions than their amphibian ancestors and during the Mesozoic the number of species increased dramatically. For this reason the Mesozoic is often called the 'age of reptiles'.

Reptiles were the dominant form of terrestrial vertebrate life for around 120 million years. Not that all species were terrestrial – many, notably the fish-like ichthyosaurs, returned to an aquatic existence, while the bird-like pterosaurs took to the air. Undoubtedly the best-known forms flourishing at this time were the dinosaurs. These huge reptiles belonged to a large and important group, the Archosauria, which also included the crocodilians (some of which have survived, little-changed, to the present day), the pterosaurs and the ancestors of birds.

At the end of the Mesozoic most of the reptilian orders died out. This is usually attributed to changes in topography and climate. It seems to have become colder and drier, which in turn led to changes in vegetation. Presumably the forms that died out were unable to adapt to these changing conditions.

Phylum Chordata
Sub-phylum Vertebrata
Class Amphibia
Order Anura

Rana temporaria
Common frog

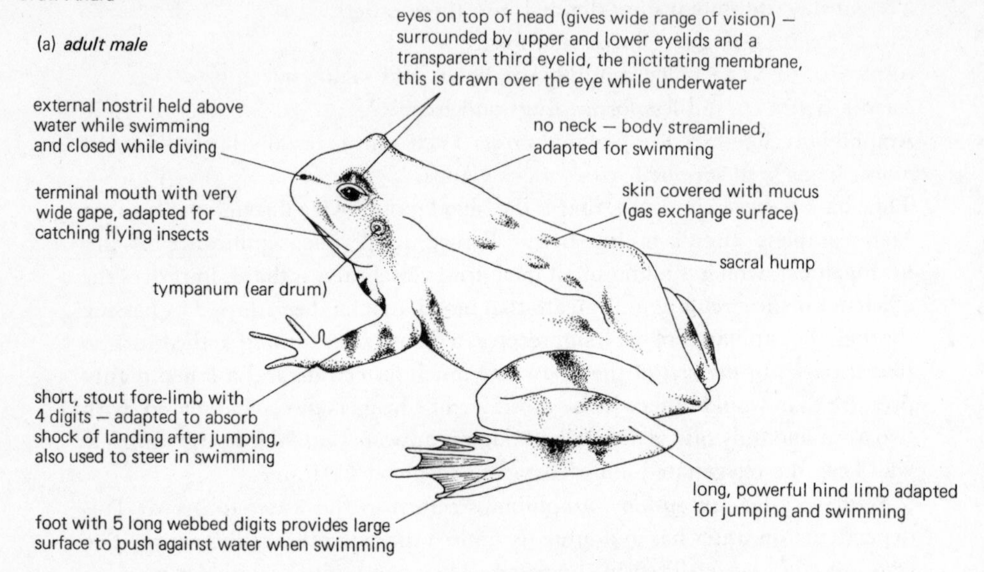

(b) *Ventral surface of hand – male frog*

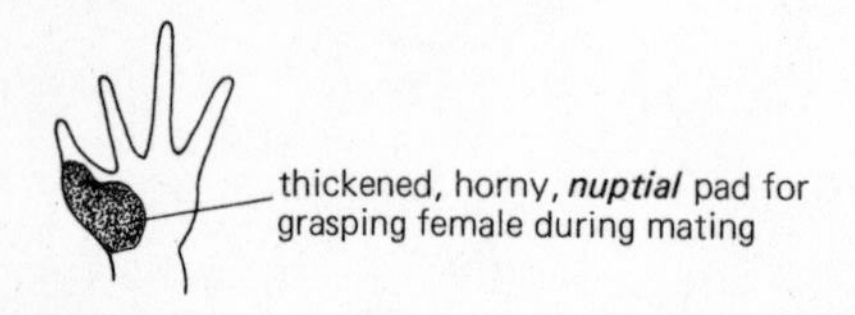

Figure 105 *Rana temporaria*, the common frog lives in damp vegetation usually near small bodies of freshwater such as ponds and ditches. It is widely distributed although it is no longer as common as it used to be, possibly as a result of the destruction of many of its breeding sites.

Frogs feed mainly on slugs, snails and insects which are caught by the rapid ejection of the frog's long, sticky, forked tongue. The tongue is attached at the front of the mouth. The upper jaw is toothed. The teeth are backwardly-pointing and help prevent prey escaping once in the mouth. Frogs can change colour slowly in response to different atmospheric conditions and different background colouring. In dry, light or hot conditions they become lighter in colour and in moist, dark or cool conditions their skin darkens.

Frogs hibernate in the winter, either under stones or in holes or in the mud at the bottom of ponds and ditches.

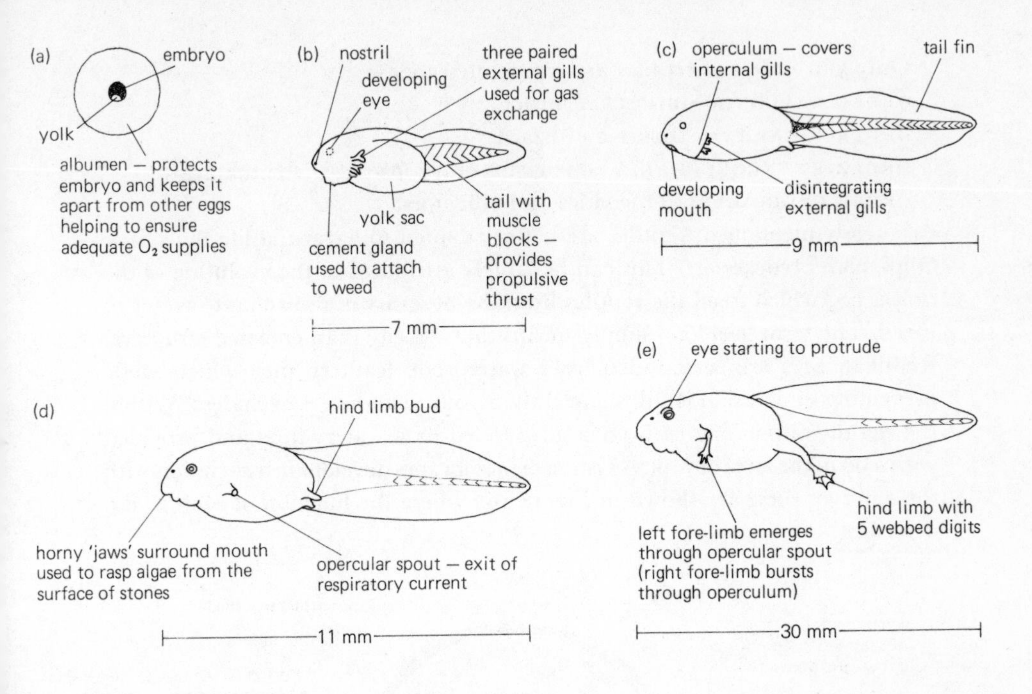

Figure 106 Stages in the metamorphosis of *Rana temporaria* (a) fertilised egg, (b) newly hatched tadpole, (c) tadpole – a few days after hatching, (d) tadpole – several weeks after hatching, and (e) tadpole – eight weeks after hatching.

Adults emerge from hibernation early in the year and make their way to a suitable breeding place such as a pond or a ditch. The male mounts the female and the two move around together for several days until the eggs are laid and the sperm shed over them. The tadpole larva emerges from the 'jelly' about twelve days after fertilisation. It attaches to weed by means of the cement gland on the underside of the head. It respires using its external gills and feeds on the contents of its yolk sac while the mouth is developing. Once the mouth has opened, the tadpole frees itself and swims around feeding on algae which it rasps from the surface of stones etc. using the horny outgrowths that surround the mouth. The operculum grows back over the external gills. The latter shrivel while four pairs of branchial clefts open and develop into internal gills. Water is taken in through the mouth and passed out over the gills and through the opercular spout. The tadpole continues to feed on vegetation and develops a long coiled intestine which can be seen through the skin. Apart from an increase in size, little change is apparent until about eight weeks after hatching, when hind limb buds start to appear. The fore-limbs appear soon afterwards. The tadpole starts coming to the surface using its newly developed lungs. The gut becomes shorter and the skeleton starts to ossify. The horny teeth are cast and the tadpole stops feeding. The ciliated epidermis is then shed and replaced by the adult skin. The eyes become prominent, the mouth widens, the tongue enlarges and the young frog adopts the adult diet.The tail is gradually reabsorbed, completing the metamorphosis. The whole process of development takes about three months, although the exact timing depends on the temperature.

Only four orders of reptiles are represented today:

ORDER CHELONIA – turtles and tortoises;

ORDER SQUAMATA – snakes and lizards;

ORDER RHYNCOCEPHALIA – one genus only, *Sphenodon* (the tuatara);

ORDER CROCODILIA – crocodiles and alligators.

As already mentioned, reptiles are better adapted to terrestrial life than their amphibian predecessors. This can be largely attributed to the evolution of the *cleidoic* egg which freed the reptiles from the necessity of returning to water to breed. The term 'cleidoic' simply means that the egg is an enclosed structure. Reptilian eggs are surrounded by a waterproof, leathery shell which while preventing desiccation is still sufficiently porous to allow gas exchange. Within the egg the developing embryo is surrounded by a watery fluid and provided with food in the form of yolk. A system of membranes develops in association with the embryo; these are shown in Figure 107 where the function of each is also described.

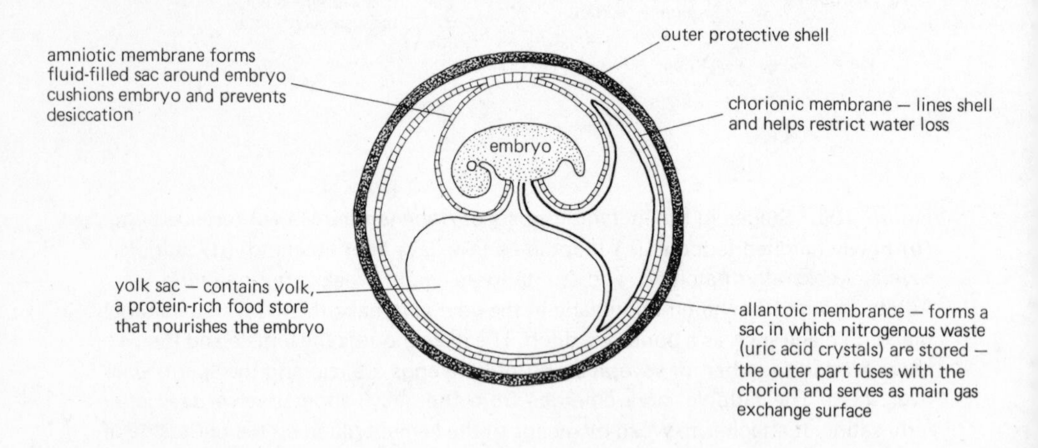

Figure 107 Diagram of a generalised cleidoic egg

Fertilisation in reptiles, as in all terrestrial animals, is internal. There is no metamorphosis, the young hatch out as miniature replicas of the adults. In many species the fertilised eggs are retained within the oviducts and the female gives birth to live young.

The skin of reptiles is also well suited to terrestrial life. It is heavily keratinised, usually in the form of closely fitting scales, providing a very efficient, waterproofing coat.

Reptiles are poikilotherms, although most species maintain a remarkably constant temperature by behavioural means. Circulation is double and the heart has two atria and a partially divided ventricle.

The brain has the same basic plan as that of fish and amphibia although it is relatively larger and there is a cerebral cortex which is not present in the brain of more primitive vertebrates.

Phylum Chordata
Sub-phylum Vertebrata
Class Reptilia
Order Squamata

Lacerta vivipara

Viviparous lizard

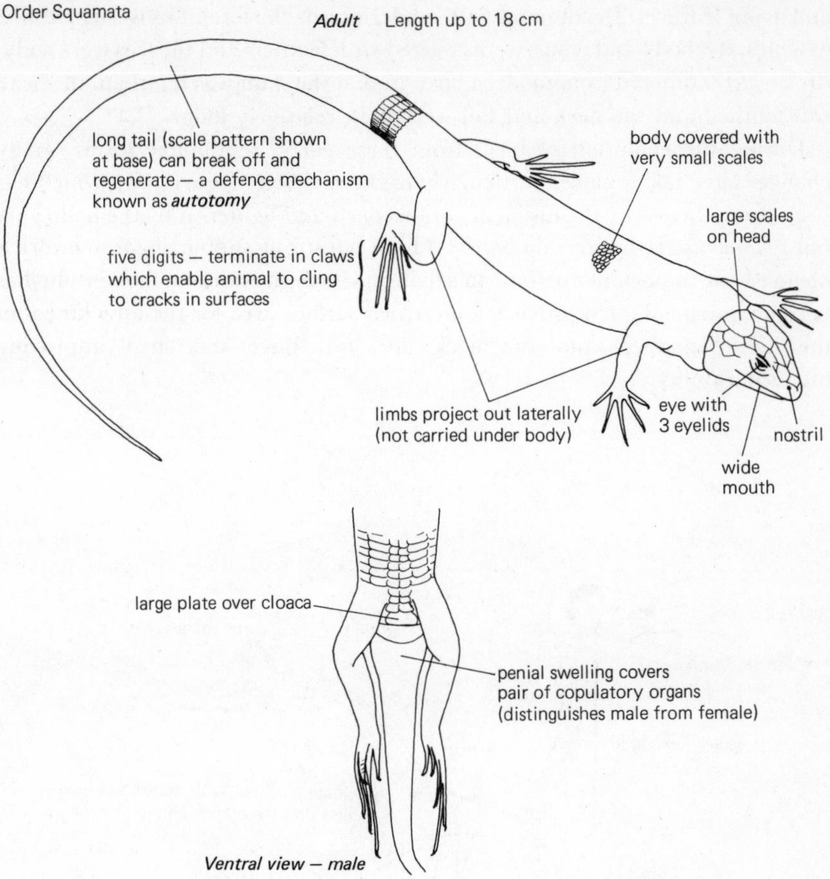

Figure 108 *Lacerta vivipara*, the common or viviparous lizard, is widely distributed throughout the British Isles. It lives in a variety of habitats including sand dunes and mountainous areas. They feed mainly on arachnids, insects and insect larvae. During the breeding season lizards tend to live in large colonies. Mating occurs in April and May. Fertilisation is internal. The fertilised eggs are retained in the oviducts of the female for about three months. The young are born alive enclosed in a thin membrane from which they soon break free and start feeding themselves.

The structure of the common or viviparous lizard, *Lacerta viridis*, is shown in Figure 108.

Class Aves

Birds first appeared in the Jurassic period. They are thought to have evolved from a group of small, bipedal, archosaurian reptiles. The earliest bird fossils discovered so far, belong to the genus *Archaeopteryx*. Although usually described as a bird, *Archaeopteryx* was in fact an intermediate form, exhibiting both reptilian and avian features. In common with modern birds the forelimbs were modified as wings; the body and wings were covered with feathers and the legs were scaly. *Archaeopteryx* differed from modern birds in that the skull was reptilian, the beak was toothed and the neck and tail were both relatively long.

During the evolution of birds from these early, reptile-like forms, many changes have taken place, particularly in the skeleton (Figure 109), which has been strengthened by the fusion of various parts and lightened by the hollowing out (*pneumatisation*) of certain bones. A light but strong supporting framework is obviously an important attribute in a flying animal. In addition, the sternum has become much enlarged providing increased surface area for the attachment of the flight muscles, and the neck and tail have shortened improving manoeuvrability.

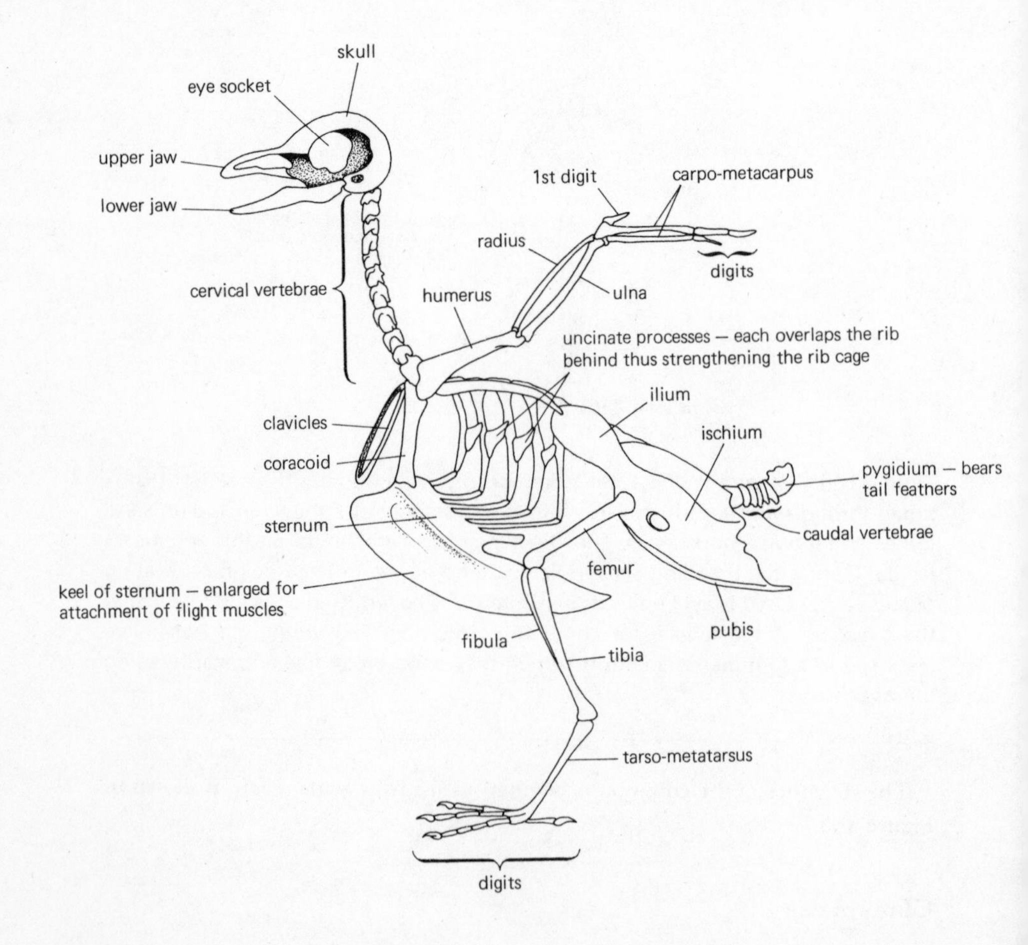

Figure 109 Skeleton of a pigeon

Birds are homoiotherms. Most maintain a body temperature between 40 and 45°C. Flight involves a considerable expenditure of energy and so necessitates a high metabolic rate. It is therefore probable that the development of homoiothermy was an important pre-requisite to the evolution of flight. Certainly the presence of feathers on *Archaeopteryx* suggests that it was a homeotherm, as feathers are thought to have evolved primarily as an insulating layer. It has recently been suggested that homoiothermy was developed by many of the archosaurian relatives of birds, notably the pterosaurs and dinosaurs.

Feathers perform a variety of functions apart from insulation. The wing feathers are particularly important in determining the shape of the wing and thus its aerodynamic properties (Figure 110).

Although feathers differ from species to species a few basic types can be recognised; these are illustrated in Figure 111. Each feather, whatever its type is formed from a dermal papilla over which keratin is deposited. Birds keep their feathers in order by preening with their beaks, during which they apply an oily secretion from the pygial gland at the base of the tail.

The colour of feathers is produced partly by pigments and partly by reflection and diffraction. Colour patterns play an important role in many aspects of bird behaviour such as concealment, courtship and aggression.

The circulation in birds is double and the heart is four-chambered. There is an extensive system of *air-sacs* within the body cavity, some of which penetrate the hollow cavities within bones. These air-sacs connect up with the lungs and serve to increase the oxygen carrying capacity of the organism. They also lower the density of the body and possibly act as insulation.

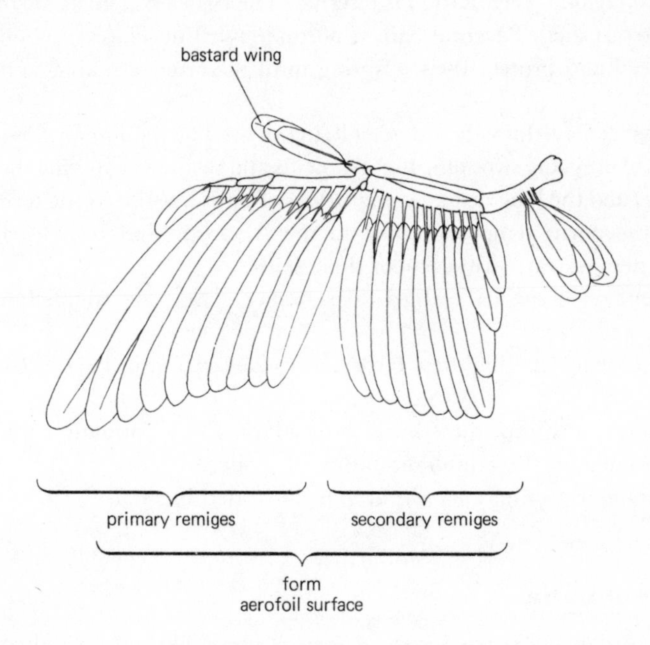

Figure 110 Arrangement of remiges on wing of pigeon

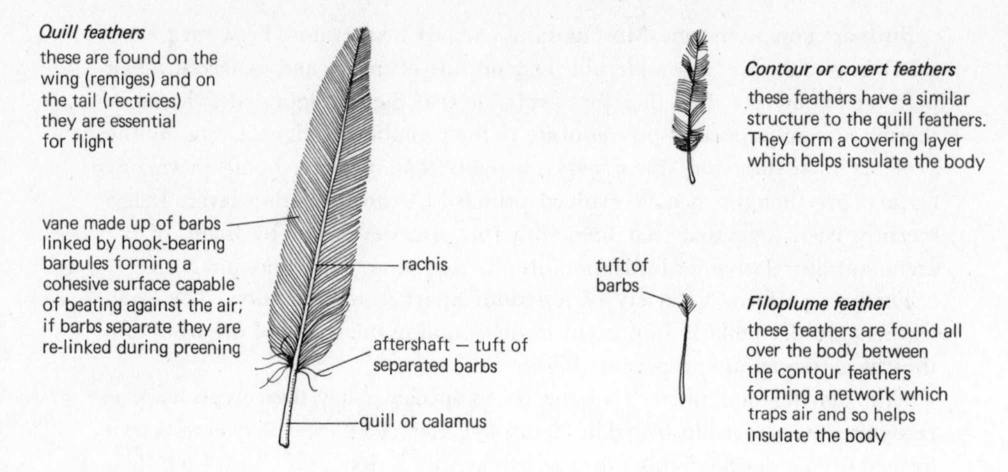

Figure 111 Types of feather

The beak and claws, like the feathers, are made of keratin. Beak shapes vary widely according to diet and feeding method.

The brain of birds differs from that of reptiles in that it is much larger relative to body size. The cerebellum and optic lobes are also much enlarged relative to other regions. The cerebellum coordinates movement – there will obviously be increased demands on this region in a flying animal. The large optic lobes reflect the dependence of birds on their eyes; sight is by far their most important sense.

The sexes are separate and pairing is usually preceded by fairly elaborate courtship behaviour. Fertilisation is internal. The eggs are cleidoic surrounded by a calcareous shell. Parental care is normally well developed; parent birds generally feed and protect their offspring until the latter are able to fend for themselves.

Birds have achieved a wide range of flying skills. The motionless hovering of many birds of prey, the swooping flight of swifts and swallows catching their prey on the wing, and the long distance migrations of birds like the Arctic tern which flies from the north temperate zone to the Antarctic and back each year, admirably demonstrate just a few of these skills.

More than 9000 species of birds exist today. These are divided into the following super-orders:

1 Palaeognathae – the flightless birds e.g. cassowary, emu, kiwi, ostrich and rhea;
2 Impennae – penguins, these birds have adapted to an aquatic existence;
3 Neognathae – all the remaining orders of modern birds.

The common pigeon, *Columba livia*, is illustrated in Figure 112.

Class Mammalia

Mammals are believed to have evolved from mammal-like reptiles (Therapsida) which lived about 180 million years ago and which showed similarities with the

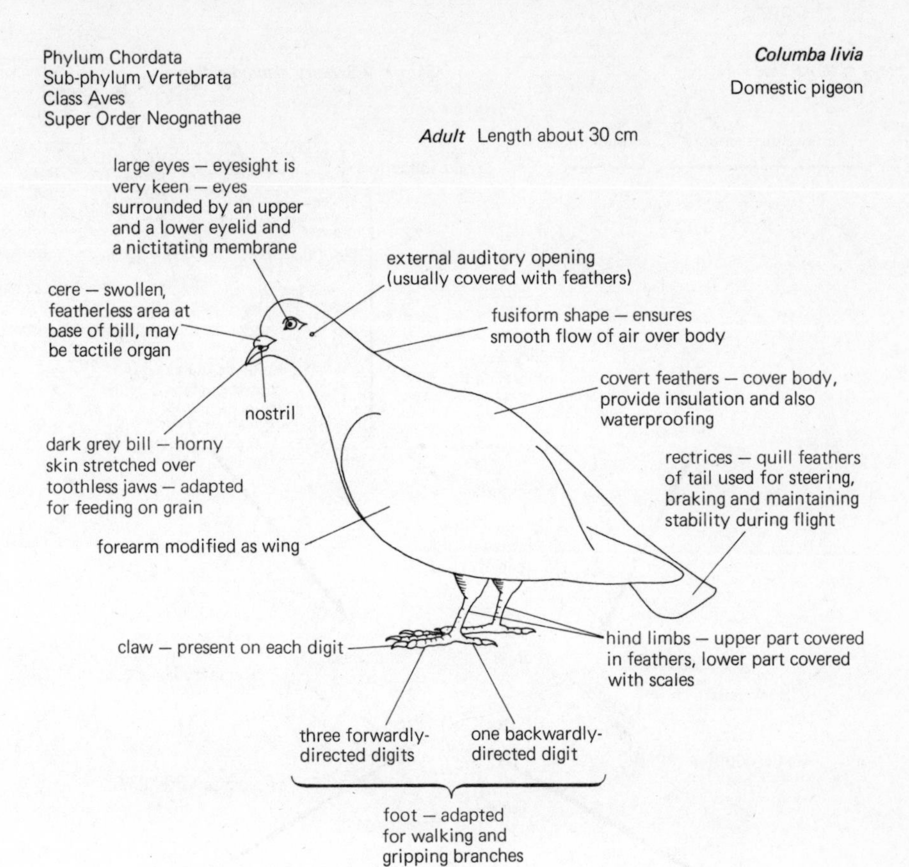

Figure 112 *Columba livia*, the domestic pigeon, is a variety of the wild rock dove. The latter is restricted to a few coastal regions of the British Isles. All pigeons feed on seeds, young green shoots and occasionally insects and snails. The nest is formed of twigs and is found on branches or ledges. The eggs are laid in spring or summer and hatch after about eighteen days. The young are fed by the mother on 'pigeon's milk', a white secretion regurgitated from the mother's crop. Pigeons are strong fliers and have been known to reach speeds of 60 m.p.h (95 km/h).

---

mammals in tooth structure, jaw articulation and limb girdles. With only bones and teeth to refer to we cannot say whether their physiology resembled that of a modern mammal. Nor can we say exactly when the first true mammals appeared, but during the last 65 million years the mammals have evolved to occupy habitats all over the world, from the Arctic to the hot deserts. There are different species which swim, burrow, fly or climb trees, showing great adaptive radiation.

Typically mammals are four-limbed, hairy, warm-blooded animals which feed their young on milk. They include many of the more familiar animals

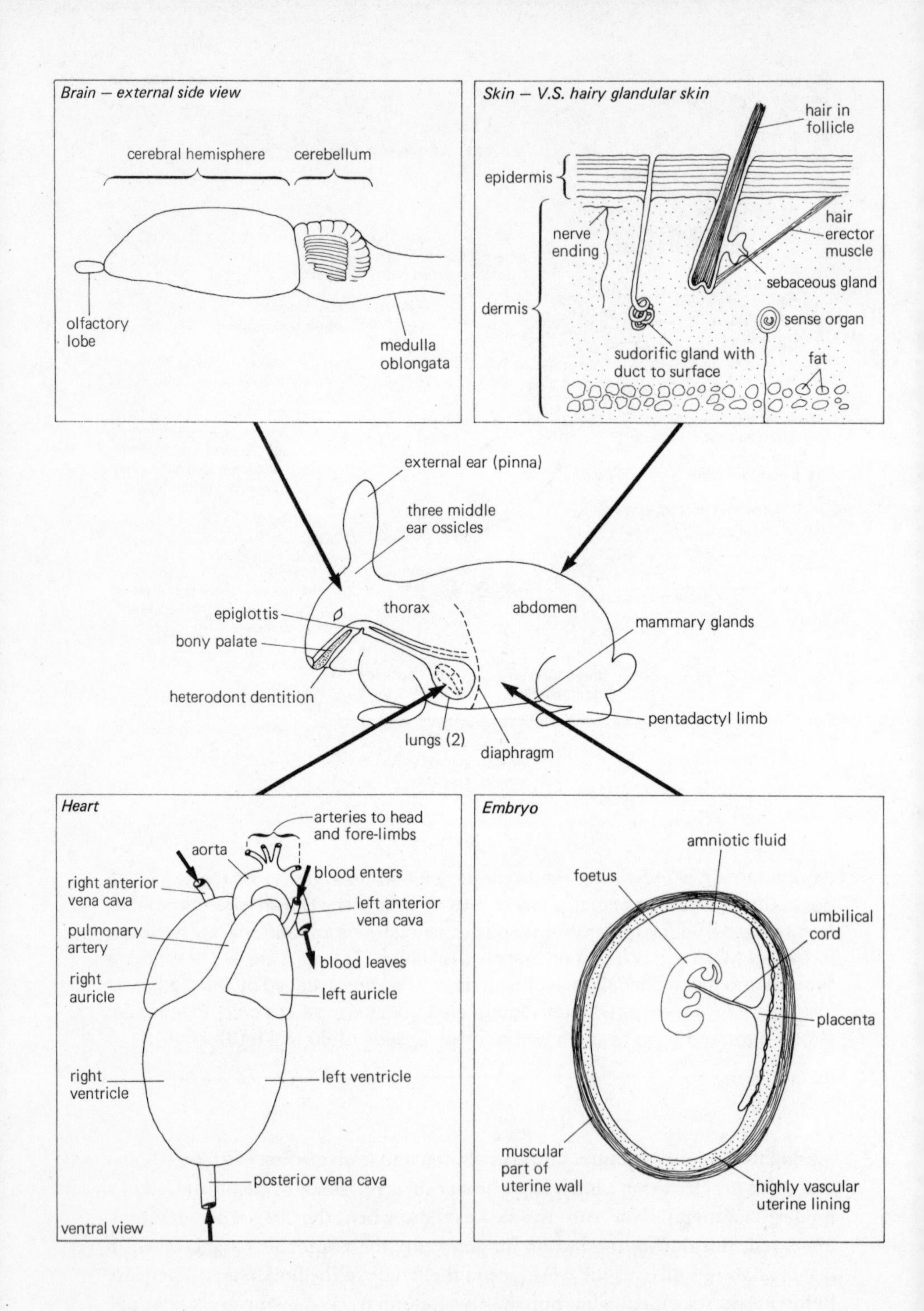

Figure 113 Mammalian characteristics

around us such as cats, dogs, cows and sheep, but also whales, elephants, lions, apes and ourselves.

Being homeothermic has helped the mammals to occupy a great many habitats and maintain a steady high metabolic rate irrespective of the environment. This is one aspect of *homeostasis*, the process by which the body of an animal is maintained in a steady state. Mammals show many examples of homeostasis and the degree of automatic control which has evolved over their internal organisation has allowed them to develop greater complexity and more rapid co-ordination than any other animal group. Figure 113 shows several of the structural characteristics of mammals.

Mammals have an insulating covering of fur which may be sparse in tropical or aquatic species, but thick in polar species. It may be prickly to deter predators and is often camouflaged by its colour or pattern. The skin also contains a layer of fat which acts as insulation. There are two kinds of glands in the skin. The *sebaceous gland* opens into the *follicle* or tube in which a hair develops and produces an oily secretion called *sebum*, which keeps the hair supple and waterproof. *Sudorific glands* excrete *sweat* – a watery solution of mineral salts and other waste products extracted from blood capillaries. These sweat glands also help to maintain the body temperature by heat loss through evaporation of the water.

The mammal skeleton has four pentadactyl limbs, but it is very common for some fingers and toes to have been lost as modifications for a particular way of life, e.g. flying or running. Unlike the limbs of reptiles, those of mammals are held under the body, raising it from the ground and allowing faster movement. It is often possible to judge the type of locomotion carried out by a limb, by observing how much of the foot is in contact with the ground. In the *plantigrade* condition the whole foot touches the ground, as in bears, apes and man, but in the *digitigrade* condition the foot has been raised so that contact is made by the underside of the fingers and toes, as in cats and dogs. This effectively increases the length of the legs, allowing faster and more silent running. In the *unguligrade* state the lengthening of the leg has continued so that the animal runs on its fingernails and toenails as in the hoofed animals. The digitigrade state usually involves a reduction in the number of toes from five to four, while in the unguligrade condition the number of digits may be three (rhinoceros), two (camel) or one (horse).

The bones of mammals have three centres of bone deposition (*ossification*), one at each end and in the centre, so that there is a relatively long period for bone growth in the animal's youth before these regions all fuse together. Unlike other vertebrates each half of the lower jaw of a mammal is composed of a single bone – the *dentary*.

Mammals are *diphyodont*, developing two sets of teeth, milk teeth followed later by the permanent set. The teeth of amphibia and reptiles are all the same – *homodont*, but mammal teeth are of several kinds – *heterodont*. At the front of a mammal's jaws are the chisel-shaped *incisors*, to the sides are pointed, fang-like *canines* and then the *premolars* and *molars*, the cheek teeth, which are used for chewing. The number and variety of teeth is a good indicator of an animal's diet.

Carnivores have large canines and pointed cheek teeth for killing and cutting flesh; whereas herbivores often have no canines and their cheek teeth are ridged for grinding vegetation. Carnivore teeth are usually arranged so that the last upper premolar cuts down outside the first lower molar in a scissor-like action; these are the *carnassial teeth* used for slicing flesh from bones. In herbivores the absence of canine teeth may result in a gap called the *diastema*, separating newly cropped vegetation at the front of the mouth from that being chewed at the back of the mouth. Because of the tough nature of vegetable foods, herbivores generally have teeth which continue to grow and be worn down throughout the life of the animal. Such teeth are called *persistent pulps*. The surfaces of their cheek teeth are generally composed of ridges of enamel and valleys of *dentine*, a softer material which wears away faster. Because the enamel always projects as ridges the teeth are effectively self-sharpening in use.

The number and type of teeth can be shown in a *dental formula* where i represents incisor, c – canine, pm – premolar, m – molar and the number present is written above a line for the upper jaw and below a line for the lower jaw. Only one side of the mouth is written down so the number of teeth must be doubled to give the total dentition.

e.g. human dental formula: $i\frac{2}{2}\ c\frac{1}{1}\ pm\frac{2}{2}\ m\frac{3}{3} = 32$

The internal organisation of the mammalian body includes the possession of *lungs* for gas exchange and a muscular *diaphragm* which separates the thorax and abdomen, and by its contraction, together with movements of the ribs, makes air move in and out of the lungs.

Separating the air passage from the food passage is a bony *palate* which forms a hard roof to the mouth and allows air to enter the nasal passages while food is retained in the mouth for chewing. When the food is swallowed a flap called the *epiglottis* closes off the air passage to the lungs (the *trachea*) and ensures that the food passes down the food tube (*oesophagus*) to the stomach.

The circulation is double (Figure 101), with the four-chambered heart lying in the thoracic cavity between the lungs.

The brain shows great development of the *cerebral hemispheres* compared with those of the other vertebrate groups, indicating better co-ordination, memory and more intelligent behaviour. Some species of mammal have complex social behaviour and even exhibit co-operative effort e.g. when hunting as a pack. The well-developed *cerebellum* also suggests improved muscle control, posture and locomotion.

Most terrestrial mammals have prominent external ear flaps, (*pinnae*), while the middle ear, which is air-filled, contains three tiny bones (*ossicles*) which carry sound vibrations from the ear drum to the sensory cells of the inner ear.

The word Mammalia refers to the *mammary* or *milk-producing glands* of female mammals, which usually lead to teats or nipples from which the young mammals can suck after birth. One group of mammals is egg-laying and another produces live young at a very immature stage, which then complete their development inside a pouch, supplied with a mammary gland and teat.

Most mammals however, are *viviparous* – they give birth to live young which have developed inside the mother's womb or *uterus*. There they grow protected from the outside world, connected to the mother by a *placenta* through which food and waste materials can be exchanged between the blood of the *foetus* (developing embryo) and the blood of the mother. After birth the young mammal is cared for by its parent(s) and during this time it may be taught skills which are important for its survival.

Classification

The mammals are divided into three sub-classes according to details of their reproduction. The Prototheria or egg-laying mammals; the Metatheria which give birth to immature young that complete their development inside a pouch or *marsupium* and the Eutheria or placental mammals, in which the young develop inside the body of the mother, nourished by her blood via a placenta.

Sub-class	*Order*	
Prototheria	Monotremata	e.g. duck-billed platypus
Metatheria	Marsupialia	e.g. kangaroo
Eutheria	Insectivora	mole, hedgehog
	Dermoptera	*Colugo* (flying lemur)
	Chiroptera	bats
	Edentata	armadilloes, anteaters and sloths
	Lagomorpha	rabbit, hare
	Rodentia	rat, mouse, squirrel, beaver
	Cetacea	dolphin, whale
	Carnivora	fox, dog, cat, bear, seal
	Tubulidentata	aardvark
	Hyracoidea	*Hyrax* (conie)
	Proboscidea	elephant
	Sirenia	sea cow, manatee
	Perissodactyla	(odd-toed ungulates) horse, tapir, rhino
	Artiodactyla	(even-toed ungulates) camel, deer, cow, goat
	Primates	lemur, monkey, ape, man

Six mammals will now be considered in more detail on the following pages and are listed below. They have been chosen to illustrate the range of habitat, bodily form and way of life of the mammals.

Rabbit – herbivore (Figure 114).
Rat – omnivore (Figure 115).
Mole – insectivore, adapted for digging (Figure 116).
Bat – insectivore, adapted for flight (Figure 117).
Fox – carnivore (Figure 118).
Dolphin – carnivorous, aquatic (Figure 119).

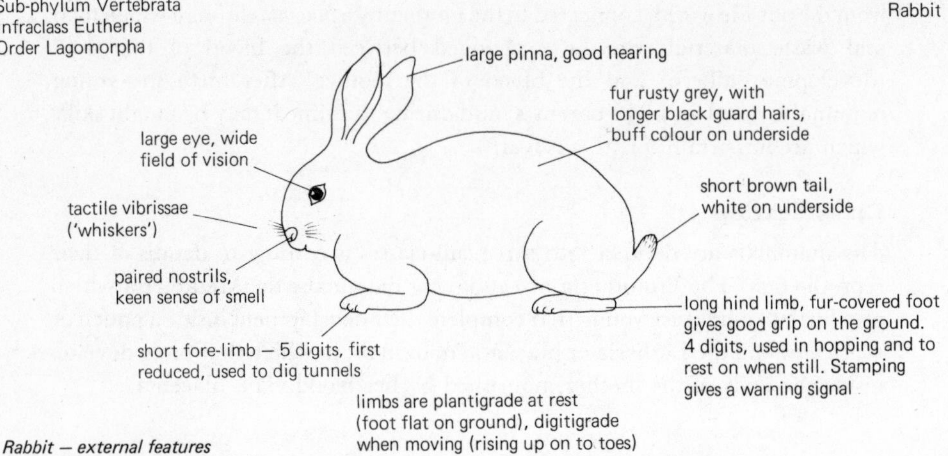

Rabbit – external features

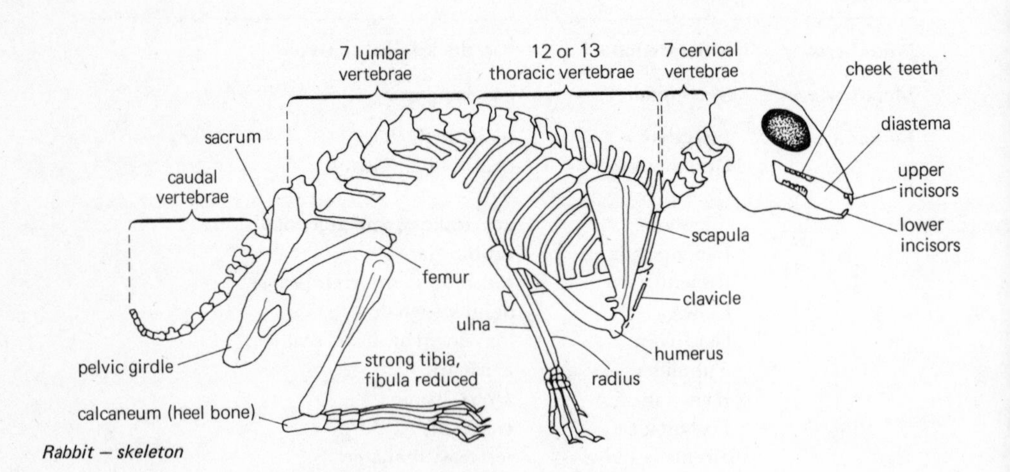

Rabbit – skeleton

Figure 114 *Oryctolagus cuniculus* – Rabbit
Habitat communities in burrows consisting of connecting tunnels with several entrances, usually dug in light, sandy soil in open woods or grassland.
Size length 34 to 45 cm, weight 1·5 to 2·5 kg.
Feeding mainly nocturnal or crepuscular, feeding on grass, seeds, leaves, young bark. They rarely drink.

$$\text{Dental formula: } i\frac{2}{1}\ c\frac{0}{0}\ pm\frac{3}{2}\ m\frac{3}{3} = 28$$

The lower incisors chew against a small pair of upper incisors immediately behind the large ones. These teeth are completely covered with enamel, unlike the incisors of rodents, which only have enamel at the front. Like the rodents', rabbit teeth grow continuously (persistent pulps) and must be continually worn down by use. A diastema separates the incisors from the cheek teeth. Rabbits reswallow the soft faecal pellets they produce at night, probably because they are a good source of

vitamin B, produced by intestinal bacteria. Suckling young will die if prevented from swallowing maternal soft pellets.
Reproduction three to six litters are produced each year between March and September. Gestation takes about twenty-eight days. There are usually 4–16 young in a litter and their eyes open at ten days. They are suckled for about four weeks and mature at around six to eight months. Their life span is up to ten years.
Predators fox, birds of prey, man.

Their very high reproductive rate, resulting from short gestation, many litters and early sexual maturity sometimes makes rabbits a pest on agricultural land. The virus disease myxomatosis was introduced to control them, but some rabbits are now immune to this disease.

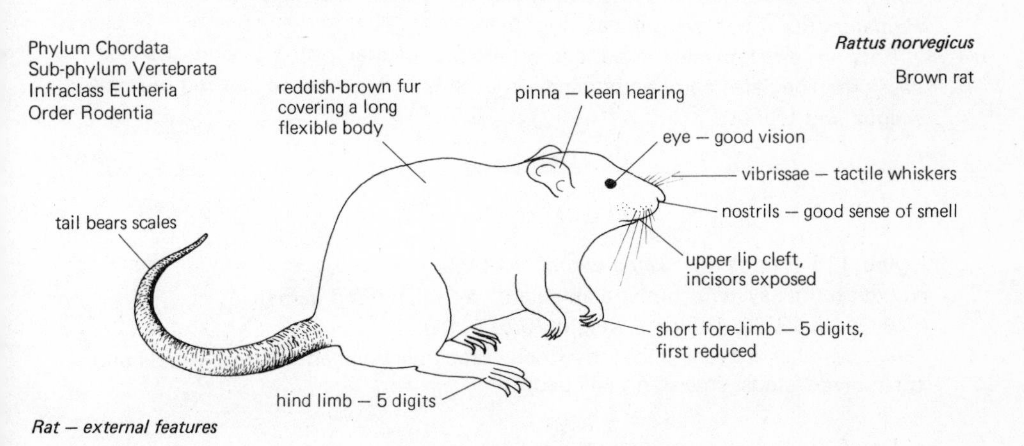

Rat — external features

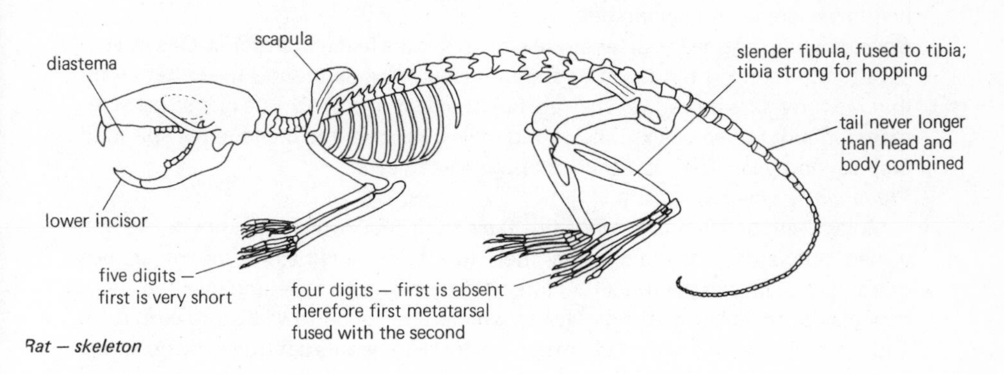

Rat — skeleton

Figure 115 *Rattus Norvegicus* – Brown rat
Habitat rat colonies develop in burrows in refuse tips, sewers and warehouses.
Size length about 40 cm including tail, weight 275 to 580 g.
Feeding omnivores, they feed on all kinds of waste food, especially grain, but also attack inedible things like wood or even water pipes. Constant gnawing is necessary because their incisor teeth grow continuously. They drink freely and can swim. (continued overleaf)

Figure 115 (cont.)

Dental formula: $i\frac{1}{1}\ c\frac{0}{0}\ pm\frac{0}{0}\ m\frac{3}{3} = 16$

The incisors are only enamelled on the outer side, which is therefore tougher, so gnawing makes the teeth chisel-shaped as the softer inside wears away faster. There are no canines or premolars, leaving a diastema. Skin can be folded inwards here and stops the food being gnawed passing back into the mouth. The molars are self-sharpening because of differential wear resulting in ridges of enamel and troughs of softer dentine.

Reproduction two to seven litters are produced a year after a gestation period of three weeks. There are four to ten naked, blind young in a litter. They are suckled for about three weeks. Young rats become sexually mature at about three months.

Predators fox, stoat, weasel, cat, dog, man.

Rats are economically important as spoilers of man's stored food and possessions. They also carry pathogenic organisms, such as those causing bubonic plague and typhus.

Figure 116 (opposite) *Talpa europaea* – Mole

Habitat tunnel systems under arable land, woodland and gardens.

Size length 127 to 165 mm, weight 70 to 110 g.

Feeding highly active animals, they feed voraciously on earthworms, insects and their larvae, slugs, snails and millipedes.

Dental formula: $i\frac{3}{3}\ c\frac{1}{1}\ pm\frac{4}{4}\ m\frac{3}{3} = 44$

The incisors are pointed for seizing prey. The upper canine is long and pointed for killing, but the lower one is reduced. The cheek teeth all have pointed cusps and the first lower premolar is elongated.

Reproduction outside the breeding season moles are fiercely territorial. One litter of about four young is produced in spring inside an underground nest after about thirty to forty days gestation. They are born hairless and blind. The female cares for them for about three weeks. The young moles will begin to breed in the spring of their second year. Their life span is about three years.

Predators tawny owl, fox, man.

Moles walk on the inside edges of their front feet, which are shovel-like and turned outwards. They dig using their fore-limbs alternately, the other legs gripping the sides of the tunnel. At intervals, by putting a fore-limb over the snout they may push soil up to the surface as a molehill. The burrows disturb crop roots and the molehills may damage farm machinery so moles are controlled by poisoned bait and traps.

Phylum Chordata
Sub-phylum Vertebrata
Infraclass Eutheria
Order Insectivora

Talpa europaea

Mole

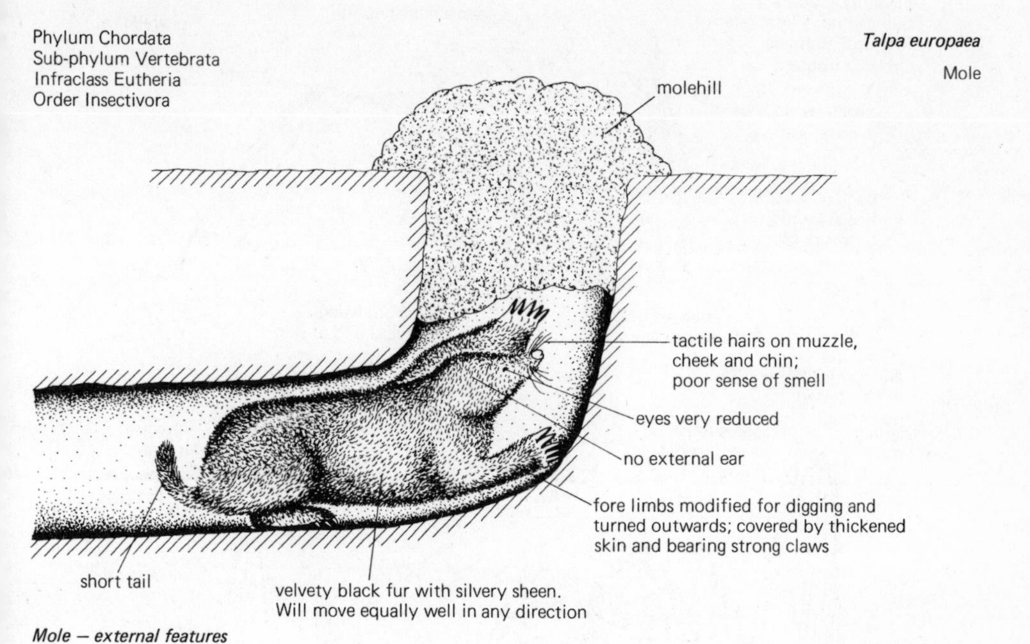

Mole – external features

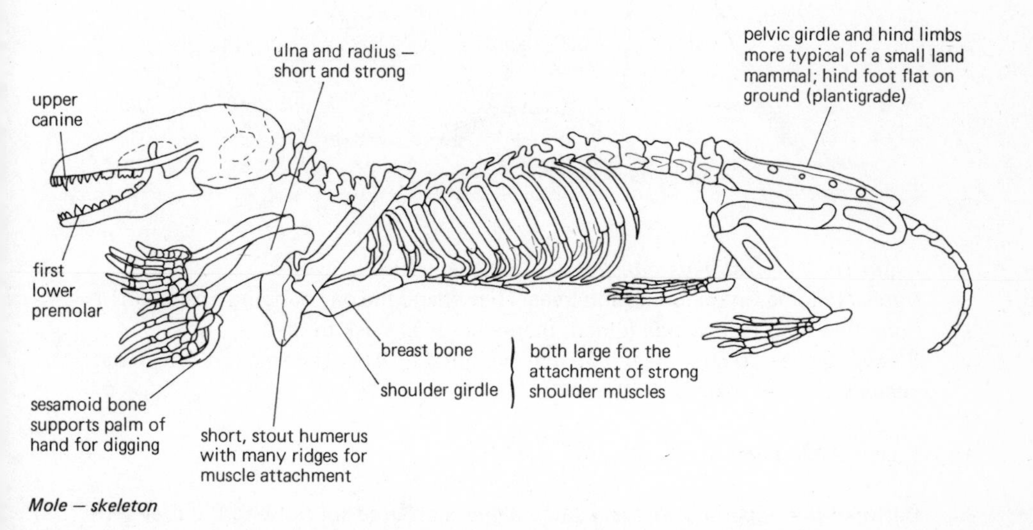

Mole – skeleton

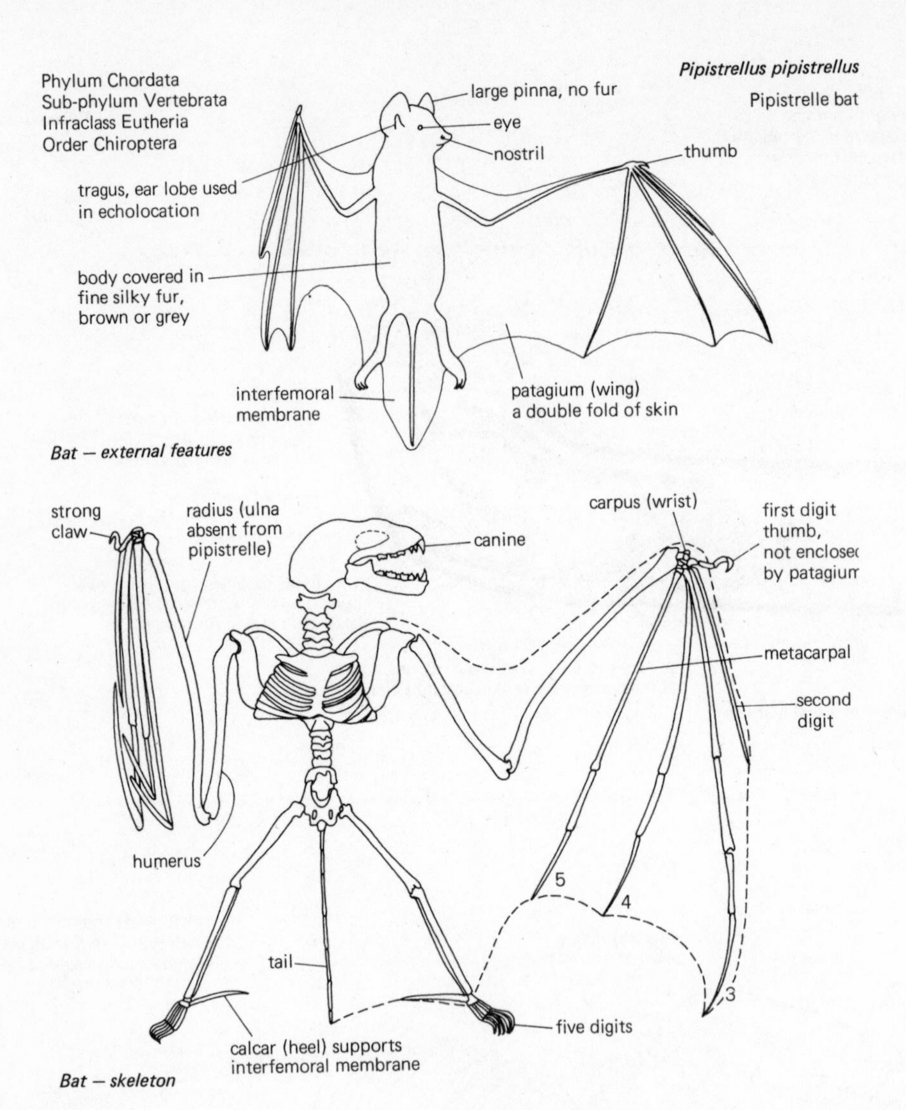

Figure 117 *Pipistrellus pipistrellus* – Pipistrelle bat
Habitat colonies in buildings and trees. They spend the day hanging upside down from their hind feet, wings folded, then emerge at dusk to feed.
Size wing span 20 cm, forearm 25 mm, weight 8 to 10 g. The smallest British bat.
Feeding catches insects on the wing.

Dental formula: $i\frac{2}{3}\ c\frac{1}{1}\ pm\frac{2}{2}\ m\frac{3}{3} = 34$

Between the upper incisors is a small depression used for holding the prey. The canines are sharp for piercing insects, while the cheek teeth have pointed cusps for crushing them.
Reproduction bats mate in the autumn, just before their hibernation, which lasts from October to March. Fertilisation is delayed until the female reawakes. The gestation period is two months. A single young bat is born, naked and blind. It is

carried on the mother's body for about six weeks and then, when too big to carry, it is left behind in the cave while she feeds. The young bats are fully grown at about four months. Their life span is up to ten years.
Predators probably no natural enemies.

Most bats are able to manoeuvre in darkness using echo-location. They emit ultrasonic impulses which are reflected back from obstacles or prey to the ears. They are the only mammals to show true flapping flight.

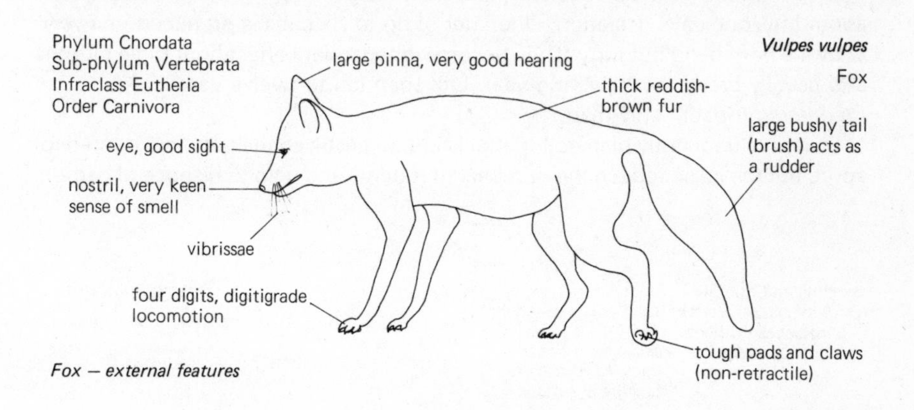

Fox – external features

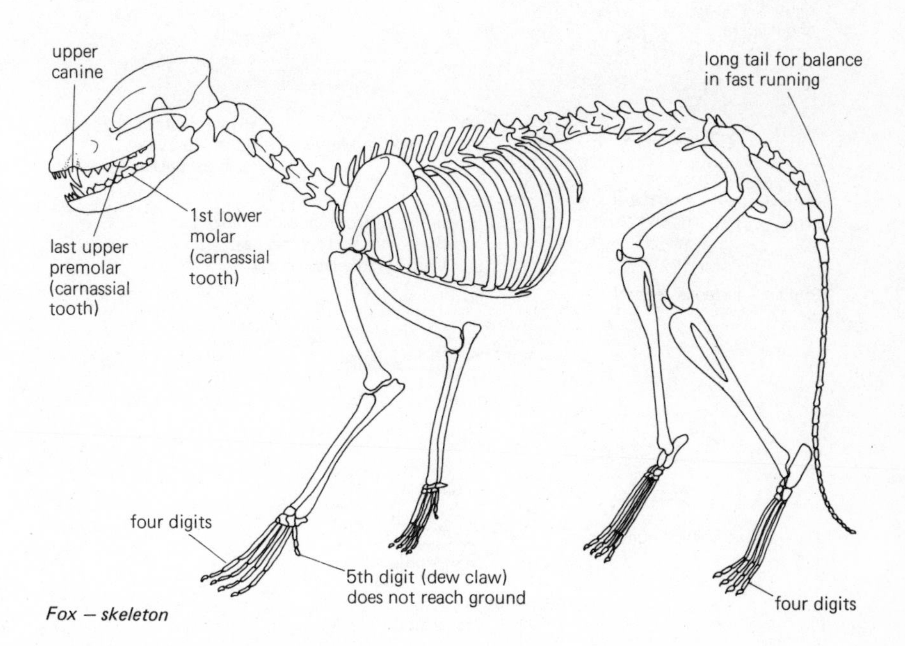

Fox – skeleton

Figure 118 *Vulpes vulpes* – Fox
Habitat fields, woods, moorland, suburban gardens. They build burrows called earths with many entrances, from which they emerge at night to feed.
Size length up to 150 cm nose to tail tip, weight 4 to 10 kg.
Feeding voles and other rodents, rabbits, small birds, beetles, fruit.
(continued overleaf)

Figure 118 (cont.)

Dental formula: $i\frac{3}{3}\, c\frac{1}{1}\, pm\frac{4}{4}\, m\frac{2}{3} = 42$

The incisors are chisel-shaped for snipping and the canines long and pointed for piercing and killing. The last upper premolar cuts down past the first lower molar like a scissor blade and is used to slice flesh. These are carnassial teeth.
Reproduction foxes mate in January and the young are born in March or April after about fifty-one days gestation. One litter of up to six cubs is produced in a year. They are born blind but furry. They become independent after about three months and usually breed the following year. Life span ten to twelve years.
Predators probably only man.

Foxes are frequently blamed for attacks on domestic animals and are hunted for sport, but they also reduce the numbers of rodents and help to dispose of carrion.

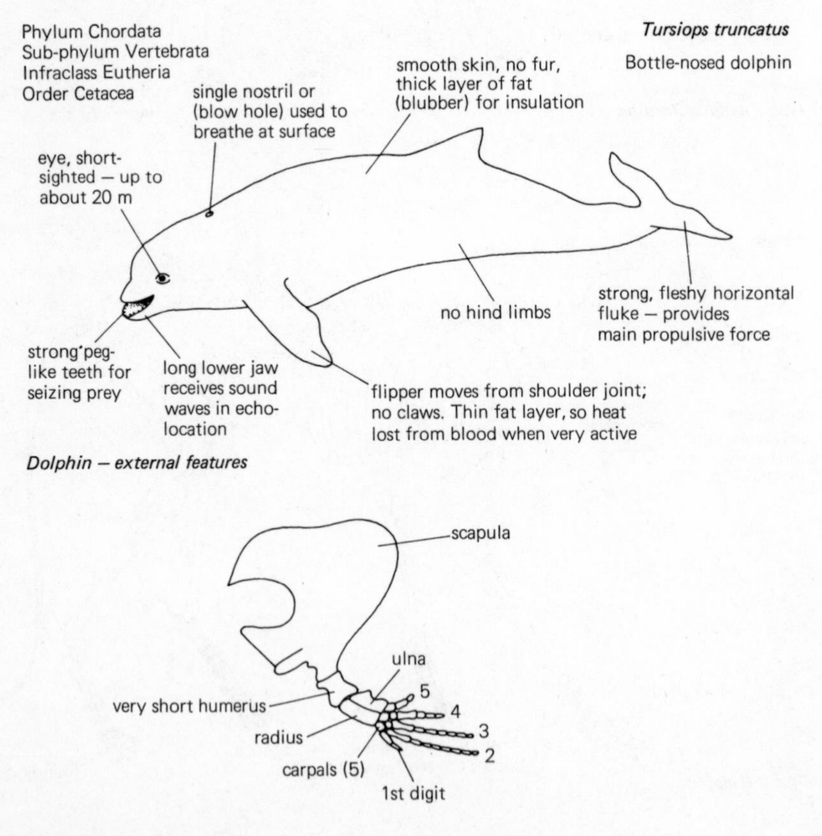

Figure 119 *Tursiops truncatus* – Bottle-nosed dolphin
Habitat North Atlantic and Mediterranean Sea. Large, stable groups with social organisation. They are active by day, swimming up to 35 km per hour. At night they rest but still have to breathe at the surface.

Size 2·5 to 3 m, weight 1·50 to 200 kg.
Feeding dolphins feed on fish in the upper layers of the sea and also on octopus and squid. The Bottle-nosed dolphin has a long upper jaw forming a 'beak'. The teeth are peg-like and used for seizing the prey which is then swallowed whole. There are twenty to twenty-six teeth in each half jaw.
Reproduction a single young dolphin is born in late summer after a gestation period of ten to eleven months. Co-operative behaviour between females has been seen at births. Lactation continues for six to fifteen months, milk being pumped into the young dolphin from the teats. Dolphins become sexually mature after five to six years.
Predators killer whale.

Dolphins use echo-location to explore their surroundings and communicate by a wide range of sounds. They are highly intelligent, with a large complex brain. They learn tricks easily and can be seen performing in dolphinaria.

QUESTIONS

1. Explain why fish, reptiles and mammals are all assigned to one phylum, the Chordata. Then list the features that distinguish each of these three classes.
2. Describe the changes that have occurred during the evolution of birds adapting them for flight.
3. Why is the appearance of the cleidoic egg seen as such a vital step in the evolution of terrestrial vertebrates?
4. Explain the meaning of the terms: homeothermic; double circulation; viviparous; homeostasis.
5. In what respects do you consider the mammals to be more evolutionarily advanced than the other vertebrates? How have these features contributed to their wide range of habitat and adaptive radiation?

19 Life cycles

A number of types of life cycle are distinguishable, depending on the relative importance of the haploid and diploid condition of the chromosomes. A haploid organism contains a single set of chromosomes; whereas a diploid one has two sets, one set being inherited from each parent. The reduction from the diploid to the haploid number is carried out by a type of nuclear division called meiosis.

Some simple organisms such as *Amoeba* remain haploid throughout their existence and reproduce only by asexual reproduction in the form of binary or multiple fission. Most organisms, however, carry out sexual reproduction at some stage in their life cycle and this requires the production of haploid cells, which then fuse and restore the diploid state. The length of the haploid and diploid phases varies and it is on this that the life cycle categories are based.

TYPES OF LIFE CYCLE

1 Haplontic

In this type of life cycle the adult organism is haploid and produces haploid gametes without the necessity for meiosis (Figure 120). The gametes then fuse to form a diploid zygote. Meiosis follows either immediately or after a resting stage, restoring the haploid condition. Any later divisions to allow growth and development into the adult condition are mitotic. Many Algae and Fungi, such as *Spirogyra* and *Mucor* show this kind of life cycle as do some Protozoa including *Monocystis* and *Plasmodium*.

2 Diplontic

Here the adult organism is diploid and meiosis is necessary to produce the haploid gametes (Figure 120). After fusion the zygote undergoes mitotic divisions to develop into the adult. Examples of this type of life cycle are found in some Fungi, Algae such as *Fucus* and the majority of animals.

3 Haplo-diplontic

This kind of life cycle has alternating haploid and diploid phases (Figure 120). A diploid organism produces haploid spores as a result of meiotic divisions. Each spore may develop into a haploid organism which then produces gametes mitotically. When gametes from two individuals fuse the diploid condition is

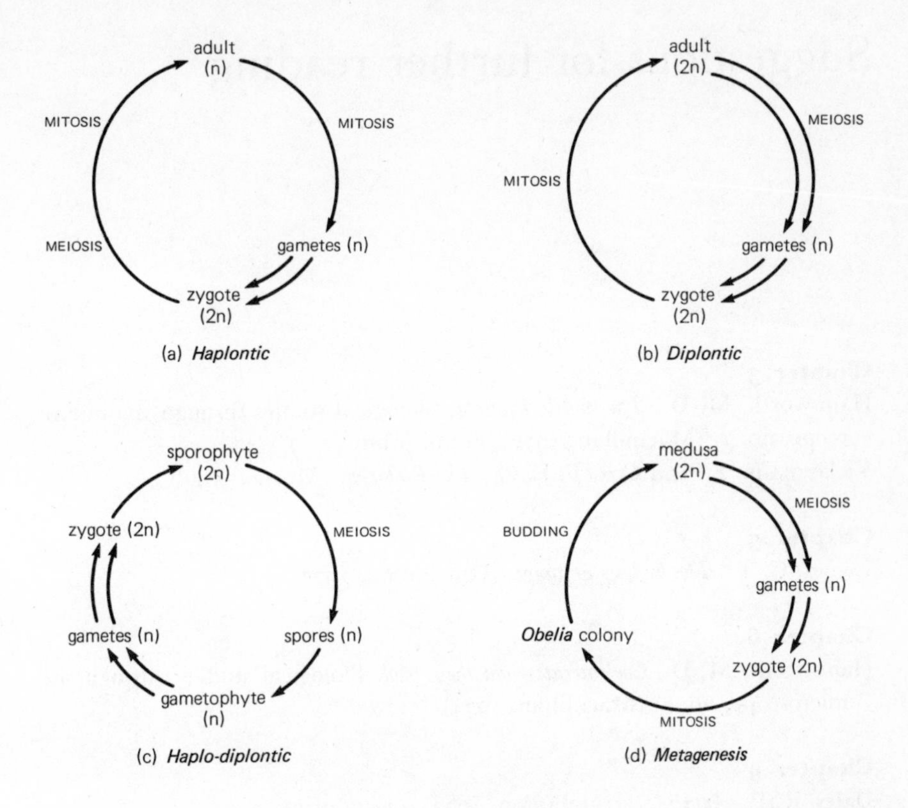

Figure 120 Types of life cycle (a) haplontic; (b) diplontic; (c) haplodiplontic; (d) metagenesis

restored. If the haploid and diploid phases are clearly separate from each other the cycle shows alternation of generations. The haploid generation is the gametophyte and the diploid stage the sporophyte. Many Algae, some Fungi and all the higher plants illustrate this type of cycle and it is seen particularly well in the bryophytes and pteridophytes (Chapters 13 and 14).

Although true alternation of generations occurs only in plants there are certain animals in the Phylum Coelenterata in particular, such as *Obelia*, whose life cycles show superficial resemblances to it (p. 51). Here there is a colony of hydroids concerned mainly with feeding, which from time to time bud off medusae. These are responsible for reproduction and dispersal of the new individuals. This is not true alternation of generations because both the colony and the medusae are diploid. It is better referred to as metagenesis i.e. delayed sexual reproduction, in this case carried out by the medusae; the colony being incapable of gamete production (Figure 120).

Suggestions for further reading

Chapter 3

Hainsworth, M. D., *The motile Protista*, Biological studies through the microscope, no. 1, (Macmillan, 1972), out of print.

Vickermann, K. and Cox, F. E. G., *The Protozoa*, (Murray, 1967).

Chapter 5

Ingold, C. T., *The biology of fungi*, (Hutchinson, 1973).

Chapter 6

Hainsworth, M. D., *Coelenterates and their food*, Biological studies through the microscope, no. 2, (Macmillan, 1974).

Chapter 9

Dales, R. P., *Annelids*, (Hutchinson, 1963), out of print.

Chapter 10

Oldroyd, H., *Insects and their world*, (British Museum of Natural History, 1973).

Wigglesworth, V. B., *The life of insects*, (Weidenfeld and Nicolson, 1964), out of print. Also in paperback, published by Mentor Books, 1968.

Chapter 11

Morton, J. E., *Molluscs*, (Hutchinson, 1967).

Chapter 12

Clark, A. M., *Starfishes and their relations*, (British Museum of Natural History, 1968).

Nichols, D., *Echinoderms*, (Hutchinson, 1969).

Chapter 14

Sporne, K. R., *The morphology of pteridophytes*, (Hutchinson, 1966).

Chapter 16

Sporne, K. R., *The morphology of gymnosperms*, (Hutchinson, 1967).

Chapter 18

Young, J. Z., *The life of vertebrates*, (Oxford University Press, 1962).

General

Buchsbaum, R., *Animals without backbones* (2 vols.), (Pelican, 1971).

Larousse Encyclopedia of Animal Life, (Hamlyn, 1967).

Mellanby, H., *Animal life in freshwater: guide to freshwater invertebrates*, (Chapman and Hall, 1975).

Robinson, M. A. and Wiggins, J. F., *Animal types*, volumes 1 and 2, (Hutchinson, 1974).

Smith, G., *Cryptogamic botany*, volumes 1 and 2, (McGraw-Hill, 1955).

Vines, A. E. and Rees, N., *Plant and animal biology*, volume 1 (Pitman, 1972).

Plant and animal parasites

Baer, J. G., *Animal parasites*, (Weidenfeld and Nicolson, 1972).

Clegg, A. G. and Clegg, P. C., *Man against disease*, (Heinemann, 1973).

Deverall, B. J., *Fungal parasitism*, Studies in Biology Series, no. 17, (Arnold, 1969).

Hainsworth, M. D., *Invertebrate parasites and their free-living relatives*, Biological studies through the microscope, no. 3, (Macmillan, 1973).

For identification

Collins Pocket Guides to Sea Shore, Wild Flowers, Mushrooms and Toadstools, etc.

Oxford Books of Invertebrates, Insects, Vertebrates, Flowerless plants, etc.

Index

Page numbers in bold type indicate major references and the pages where terms are first defined (i.e. where they appear in italic type in the text). Page numbers in italic type refer to illustrations.